Disclaimer

Working with PTSD or other trauma is challenging, it is important to talk to you physician first before beginning any treatments. This book does not take the place of traditional therapies; it is an adjunct to be used along with present therapies to aid in your recovery.

Bodywork & PTSD by Thomas Schieffer
© Copyright 2020 Thomas P Schieffer

I dedicate this book to two great teachers that have generously shared both their knowledge and continuous support with me. Shunji Morimoto and Lonny Jarrett.
Thank you both.

I would also like to thank everyone that helped to make this book a reality for me.

Editing and amazing life coach; Jeannine Doyle
Photos; Justine Cranford
Graphic design; Kevin Smithey
Models; Abby Wales, Zach Horn

Self Publish by Thomas Schieffer
Thomas-schieffer.com
tpschieffer@gmail.com

Forward

Thomas's professional journey to become a Shiatsu therapist began in a round-about way.

Thomas enjoyed restoring cars and began flying while in high school and also cultivated his strength working with his hands. However he struggled with dyslexia and found the conventional path was never the right fit for him.

Thomas then became a Black Hawk helicopter mechanic and spent 10 years flying around the world and learning about different cultures.
He saw a window into the world that both humbled and challenged him with both the beauty and deep sadness he witnessed.

Returning to America, Thomas went on a 2 year 'walk-about' to heal his own heart and find a new life path. He realized he couldn't get away from his challenges without looking at himself. During this time he became involved with a yoga community and began working with an acupuncturist and herbalist. He started his own apothecary and herb farm while learning both Chinese and Western herbal medicine. He worked with David Crow, founder of Floracopeia, learning how to distill essential oils, make herbal medicine, and learned the business of herbalism. He also learned the deeper aspects of medicine cultivating the medicinal qualities through plant's life cycles in nature, what Thomas calls 'Qi Medicine'.

Thomas then began training with renowned herbalists, qi gong masters and acupuncturists including Lonny Jarrett, Michael Tierra, and Dr. Yang Jing Ming.

At another crossroads, he realized that making a living as an herb farmer wasn't sustainable. Thomas's path then crossed with Shunji

Morimoto, a Shiatsu Master who studied in Japan and lived the life that Thomas describes as Buddha. "He was incredibly calm and had endless amounts of energy and happiness."

Thomas studied closely with Shunji weekly for 3 years while living in an Ashram and practicing meditation.

Thomas also learned ancient techniques and body mechanics while integrating the Chinese Medicine meridian system into his shiatsu training. He learned that Shunji's style of Tao Shiatsu was cultivated thousands of years ago as a form of healing the body and mind by addressing the entire energetic balance of the body, bolstering our innate strength to heal ourselves. Today, Thomas carries Shunji's style of Tao Shiatsu forth.

Over the years Thomas refined his skills and learned more clinical approaches for traditional diagnosis that complimented his herbal knowledge.

Thomas is known for his work with people who have experienced trauma and want to naturally attune their body and mind to healthier stress responses. He also enjoys working with athletes, yogis, and anyone who uses their body for healing work.

Shunji

My teacher and mentor, Shunji Morimoto is a man of tremendous compassion, wisdom and truth.

While working with Chinese medicine (TCM), I was told of a a shiatsu master, Shunji, who had over 30 years of experience in practicing shiatsu with clients from around the country. Shunji also trained with some of the leading shiatsu masters who created Zen Shiatsu.

At the time, I didn't know what shiatsu was or how it healed the body and I was very curious. I booked an appointment. Shunji was a short man, standing just 5'5" with gray and black hair, and dressed in wild-colored pants with patches on the knees and a white shirt with "Weird Wendell" printed on it. He was wearing a yellow hand towel around his neck, toe socks so he could wear his sandals, and, to top it off, a great big smile that made me feel as if he had been waiting all his life just to see me and that I had nothing to worry about.

He performed shiatsu on me for two hours. He explained that shiatsu involved rhythmic pressing, pulling, working on the body's pressure points, and the movement of the body. Experiencing a lifetime of PTSD made me think this was just another technique that wouldn't give me relief but to my surprise, I loved it! I experienced a semi-awake state of extreme relaxation. After that first session, **I was hooked and have never looked back.**

I asked Shunji a few questions about his practice. With pure joy in his eyes and no hesitation in his voice he exclaimed, "I love shiatsu!" I was amazed at how someone could love his work so much. I knew right then that I needed to train with him. After I asked if he would teach me, I was surprised to find out that he didn't offer shiatsu training. For the next five months, I kept calling him and finally he agreed to be my instructor.

I spent almost three years apprenticing with Shunji. During this time, I formed a group of people who also wanted to learn shiatsu. It was our place of joy, and we looked forward to it every week. All of us who apprenticed with Shunji developed a tremendous love and respect for both him and shiatsu. This apprenticeship gave me both the confidence and wisdom to realize this healing work was leading me away from the dark tunnel of my life.

Training with Shunji made a major impact on my life and I carry this love of shiatsu to all my clients, students, and everyone I meet. Shiatsu is a way of life for me now. It has taught me how to not only live, but to enjoy my life. Shiatsu has also taught me another kind of strength, well said by renowned Tai Chi master Cheng Man-ch'ing, "Softness is a quality of true self, that which exists beneath our myriad defenses. Resistance is rooted in our lack of faith in the self. We create armor to protect that self from the world: hard images of straight and brittle, false fronts. These images exact a huge toll of energy required for their maintenance. Because of this psychic armor, we are blocked from creating or loving, and the awareness that we are living a lie compounds the fear and self-disgust."

TABLE OF CONTENTS

Shiatsu and its Holistic Approach to treating PTSD 1
Chapter 1 2
  Shiatsu 2
  History of Shiatsu 2
  Shiatsu & PTSD 3
Chapter 2 5
  Levels of Training 5
  Pre-Shiatsu Setup 5
  Safe Environment 6
  Stages of Approach 6-7
  Principles of Practice 7-8
Chapter 3 9
  PTSD Protocol 9
  Chinese & Western Trauma Diagnosis 9
  Protocol Sequence 10
  Protocol Points In-Depth 13-15
Chapter 4 16
  Treatment Stories 16-21
Chapter 5 22
  Pressure Techniques 22-24
  Hand Techniques 25-29
  Leg/Arm Techniques 30-31
Chapter 6 32
  Short Sequence (Chair) 32-38
Chapter 7 39
  Long Squence 39
  Seated (Mat) 39-46
  Side Lying (Mat) 46-59
  Face-Up (Mat) 60-65
Appendix A—Channel Abbreviations 66
Appendix B—Meridian Chart 66
Glossary 69-74
Reference List 75-76
Notes 77-83

Shiatsu and its Holistic Approach to treating PTSD/Trauma

Post-traumatic stress disorder (PTSD) and trauma are an all-encompassing physical and mental pathology that greatly upsets people's lives and can cause major psychological and physical problems. It reaches far past the person affected by it; it also touches immediate family members, their community and co-workers.

This book offers a piece of a larger, holistic approach to treating PTSD and trauma. There are no shortcuts to a full recovery. It takes dedication and a willingness to explore a multifaceted methodology within a supportive environment for a balanced healing approach. One size does not fit all, but with a strong circle of support, Shiatsu offers an opportunity of finding a way out of the destructive cycle that slowly affects every aspect of one's life, offering the opportunity to live a fuller and more satisfying life.

PTSD and trauma manifest differently in each person, making shiatsu a great option, playing a significant healing role with each individual. Many people suffering from PTSD or trauma have seen their symptoms resolved or dramatically decreased over time with shiatsu.

This book of is written to be accessible to multiple levels of experience in shiatsu from beginners to advanced. It is not meant to take a medical or psychologist's role in treatment; instead, its goal is to help better understand how to use the key methods of good shiatsu training, offering great value to one's life.

Chapter 1

Shiatsu

In the Japanese language Shiatsu means "finger pressure." Shiatsu includes massage with fingers, thumbs, feet, knees, and palms along with stretching and modulation of joints to access meridians and deep embedded pain. To diagnose, the practitioner uses palpations of meridians along the whole body and specifically the Hara (rib cage to hips) or pulse diagnosis. You can use a general shiatsu sequence for overall health or diagnose and treat for more specific ailments.

The Japanese Ministry of Health defines shiatsu as "a form of manipulation by thumbs, fingers, and palms without the use of instruments, mechanical or otherwise, to apply pressure to the human skin to correct internal malfunctions, promote and maintain health, and treat specific diseases. The techniques used in shiatsu include stretching, holding, and most commonly, leaning body weight into various points along key channels."

History of Shiatsu

The very beginnings of shiatsu originated in India from a form of healing called Aryvadia. As stories say that an Indian prince called Bouty Da Mo hiked over the Himalayas into Tibet then into China teaching the ways of Aryvadia. The Chinese converted Aryvadia into Traditional Chinese Medicine (TCM) about 5,000 years ago. TCM spread throughout China, taking on different aspects to fit the needs of the climate(hot-cold/wet-dry) in that area. It was then brought to Japan by a Buddhist Monk in the sixth century.

Shiatsu originated from an ancient form of massage called Anma or Tuina. Anma has been used for thousands of years in Japan and China and was generally reserved for the blind and prohibited for sighted people to practice.

Takuiro Namikoshi started the first shiatsu school in Japan in 1940, formally making shiatsu a government recognized treatment method. His approach involved incorporating western medicine in his method.
He is considered the father of shiatsu even though the term "shiatsu" has been used since 1919 in the attempt to separate shiatsu from Anma.

Shizuto Masunaga began his bodywork training with his mother, who was a master of treating and diagnosing exclusively through the Hara (abdomen).

Shizuto altered the existing shiatsu style adding more western ideas and incorporating all 12 meridians both in the hands and feet. He started a school called Loka Shiatsu Center where he taught his new version "Zen Shiatsu."

He also incorporated the use of elbow and knees where traditional shiatsu was mostly the use of finger and hand pressure.

Shiatsu & PTSD

Shiatsu is designed to help the body's innate ability to heal, recover and thrive from the stresses of life. It is especially effective in treating deep-rooted trauma that is held in the body. Shiatsu is simple but has profound and lasting results. Shiatsu works by addressing the source of imbalances in the body, which strengthens resiliency. Shiatsu can also be performed without any extra equipment or special facilities, which makes it accessible and versatile for a wide spectrum of the population with varying economic status.

The Mayo Clinic definition of PTSD

"Post-traumatic stress disorder (PTSD) is a mental health condition that's triggered by a terrifying event — either experiencing it or witnessing it. Symptoms may include flashbacks, nightmares and severe anxiety, as well as uncontrollable thoughts about the event. Most people who go through traumatic events may have temporary difficulty adjusting and coping, but with time and good self-care, they usually get better. If the symptoms get worse, last for months or even years, and interfere day-to-day functioning, may be an indication
of PTSD.

Getting effective treatment after PTSD symptoms develop can be critical to reduce symptoms and improve function."

For more info visit Mayo Clinic on-line or talk to your doctor.

Shiatsu

Chapter 2

Levels of Training

People suffering from trauma generally do not fit on a standard recovery timetable. Practicing the techniques outlined in this book are essential before integrating them effectively into your shiatsu practice. Keep an open approach and positive attitude while being focused on the protocols and how your client responds to the shiatsu techniques both in a positive and/or negative way.

Beginner. This is the basic practice for people who have a general knowledge of shiatsu. There is no diagnosis or treatment plan at this level. At the Beginner level, shiatsu can be very effective and is ideal for family members, loved ones, nurses, and alternative healers, to name a few.

Intermediate. This is the path of someone interested in diagnosing and treating clients. It is suitable for the person willing to devote a great deal of time understanding the techniques and philosophy behind shiatsu.

Advanced. This is for someone who has devoted a great deal of time working with and studying the science and art of shiatsu. The Advanced level is for someone who has chosen shiatsu as their main profession and life pursuit.

Pre-Shiatsu Setup

Safe Environment
Setting the stage of safety is important for clients suffering from PTSD. People with PTSD are hyper-aware of their surroundings. Keep in mind that everyone has different triggers that can cause a full-blown trauma episode. Therefore, the first line of questioning with a new client is how to find ways to help them feel safe. Creating a safe environment dramatically helps treatments be far more effective.

Stages of Approach

Use the following steps in providing a safe environment for your clients.

1. **Client's health and mental history.**
   Ask questions to gather a good picture of any relevant information that can apply to your client's treatment.

   Here are some sample questions:
   - Do you have a diagnosis of PTSD, and who gave it to you?
   - How long have you had PTSD?
   - What do you do to manage the symptoms?
   - Is there any area of your body that you don't like to be touched?
   - Do you currently have a treatment plan in place?
   - What are you hoping to gain from these shiatsu sessions?

2. **Personal intention.** Set your intention before every initial interview and session. Take the time to set the intention of "What can I do to help this person?"

3. **First session.**
   a. **Outline the protocol of the session:** Explain to the client everything that will happen during the session. For example, "This session consists of me pressing, squeezing, and rotating different parts of your body with my hands, elbows, and knees to help relieve tension in your body. Sessions are usually done with your eyes closed, but if you need to open them, feel free to do so."

   b. **Safety word:** Make an agreement with your client that if they feel uncomfortable in any way, a safety word can be used to help them stop the session for any amount of time. If the session is stopped, ask the client to take a few deep breaths, sit or lie down, maybe offer them a glass of water. Suggest looking around the room. If the client's symptoms become challenging, have local phone numbers of mental health professionals available or someone they may want you to contact if they feel too uncomfortable to leave the session by themselves.

4. **Touch.** Let your client know where you will be touching their body next. Before you begin shiatsu on any part of the body, apply steady, firm, slow, rhythmic pressure to alleviate the tension held in different parts of the body. Keep contact with their body and use your Mother hand as often as possible.

5. **Environment.** Work in a space that is free from interruptions and loud noises.

Consider the following:
- Solid walls and a locking door (to create a secure place with no interruptions)
- Soundproof (keep loud noises from disrupting the session)
- Visually pleasing and inviting (plants, artwork, soft colors)
- Comfortable room temperature
- Sound machine with white noise or soothing music
- Clean, well-organized, clutter-free space

Principles of Shiatsu Practice
- Always work from the center of the body to the fingertips and toes
- Apply pressure light to deep then back to light
- Always work from your Hara (center of the body)
- Always use a Mother hand with the treatment hand
- Repeat each movement three times

Instill a caring Intention. This is one of the cornerstones of shiatsu, holding the intentions of "cause no harm" and "care/compassion." With those two intentions, you will be on the right path to helping clients heal more effectively.

Always work from the center of the body to the fingertips and toes. Chinese medicine calls body energy "Qi," which is like electricity. When too much electricity gets stuck in one place, it equates to pain. Most of the time the energy needs to be led out of the body, especially from the places where it is built up. To do this, you release the pent-up energy through fingertips and toes, which are the exit points of
the body.

Apply pressure from light to deep then deep to light. The body has a defense mechanism that keeps itself safe. To counter this, start the massage shallow, slowly working deeper and accessing even deeper levels of stagnations, then lead the energy out of the body by going back from deep to light pressure.

Always work from your Hara (center of the body). When you engage your Hara center, your client will generally relax which will allow you to go deeper in the healing session.

Always use a Mother hand with the treatment hand. The Mother hand stays in one spot to help your client relax and stay connected to both your client and yourself. This is an excellent way to monitor the level of relaxation in the body.

Repeat each movement three times. Repeating three times helps ensure you are thorough in your treatment of the body.

Practice, practice, practice. The only way to further your understanding of the positive effects of shiatsu is to keep practicing. There are no shortcuts to mastery. The key mindset is to hold a childlike interest in shiatsu not childish—a curiosity of amazement and interest in what is happening, what may happen and being aware of how healing is manifesting through your client's body and energy field.

Chapter 3

PTSD Protocol

Chinese Medicine Trauma Diagnosis
Trauma is an obscurity of the heart with phlegm and the energetic disconnection of the heart and kidney.

Western Translation of Chinese Medicine Diagnosis
When trauma occurs, it overstimulates the nervous system. After the trauma, the nervous system continues to keep the body in overdrive and fills it continuously with extra energy.

Treatment Goals
The primary goal is to reconnect the heart and kidney to ground the emotions in the heart. The secondary goal is to work on the stomach-spleen meridian to help with the phlegm that is obscuring the heart. In addition, since the liver channel controls the flow of Qi, it will offset other organs if it is overworking. Most often, this means that work is also necessary on the following meridians: heart-small intestine, kidney-bladder, and liver-gallbladder. The main objective is to remove the extra energy from the nervous system giving the client a new baseline of relaxation. This also resets the nervous system allowing
it to integrate the newly corrected state.

Frequency
Treatments are most effective if sessions are consistently scheduled about a week apart until positive effects are experienced. Once that occurs, sessions can usually be scheduled every two to four weeks. (Continuing every week, however, will not adversely affect the client in any way and they may prefer this regularity.)

Note: Counterindications: If pregnant or before/after surgery consult a medical professional before treatment.

Protocol Sequence
The main objective of the protocol is to lead the pent-up energy of trauma out of the body. You will work from the center of the body (Hara) outward to the fingers and toes. The following sequence opens the important "main gates" of the body to reestablish normal circulation. When this occurs, the body will usually begin to heal itself and find its own homeostasis.

All point locations are approximate. For reference see the chart in this chapter for exact locations on pages 12, 66-68.

Upper Body Points
- GB 21
- SI 11
- LI 10
- PC 4
- LI 4 and LU 10
- PC 8 and HE 8

1. Start at GB 21 to open the head, neck, and shoulders.
2. Move to SI 11 to open the upper body, chest, and shoulders.
3. Move to LI 10 to open the arms.
4. Move to PC 4 to open the arms.
5. Move to LI 4 to LU 10 and open the hands.
6. Move to PC 8 and HE 8 to open the hand. This is the last point, and it is the most important point to press if you cannot reach the other points in the upper body.

Lower Body Points
- BL 54
- St 31
- BL 40
- ST 36
- SP 6
- BL 60
- KD 1

1. Start at BL 54 to open the hips and lower back.
2. Move to St 31 to open legs and hip.
3. Move to BL 40 to open lower back, legs, and hip.
4. Move to ST 36 to open the legs and hip.
5. Move to SP 6 to open the legs, hips, and feet.
6. Move to BL 60 to open the legs and feet.
7. Move to KD 1 and open the legs on the last point. This point grounds the body and mind and is the most important point to press for the whole body.

Channel Abbreviations

BL	Bladder
CV	Conception vessel
GB	Gallbladder
GV	Governing vessel
HE	Heart
KD	Kidney
LI	Large Intestine
LU	Lung
LV	Liver
P	Pericardium
SI	Small intestine
SP	Spleen
ST	Stomach
TW	Triple warmer

GB9
GB20
GB12
GB21

BL10
SI11
BL11
BL13
BL14
BL15
BL17
BL18
BL10
BL20
BL23
HT5
HT7
BL27
BL28
BL29
BL30
BL31
BL32
BL33
BL34
SI7
BL54
BL38
BL40
KI9
KD1
BL60
BL61

GB14
KI26
KI27
ST23
PC4
ST25
LI10
SP15
PC5
PC6
PC7
TW5
LU10
LI5
PC8
LI4
ST31
HE8
ST32
ST36
LR6
SP6
ST41

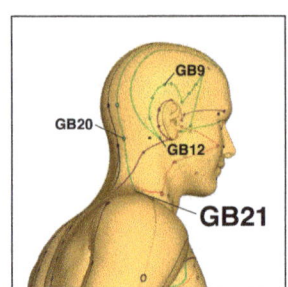

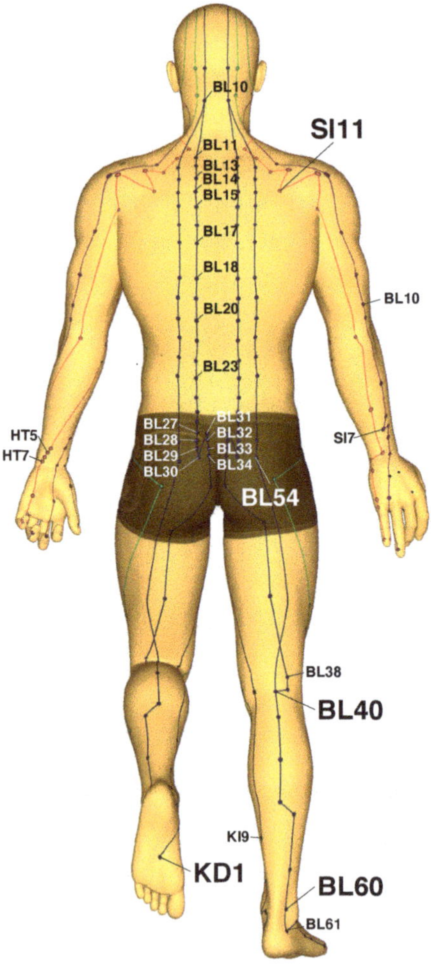

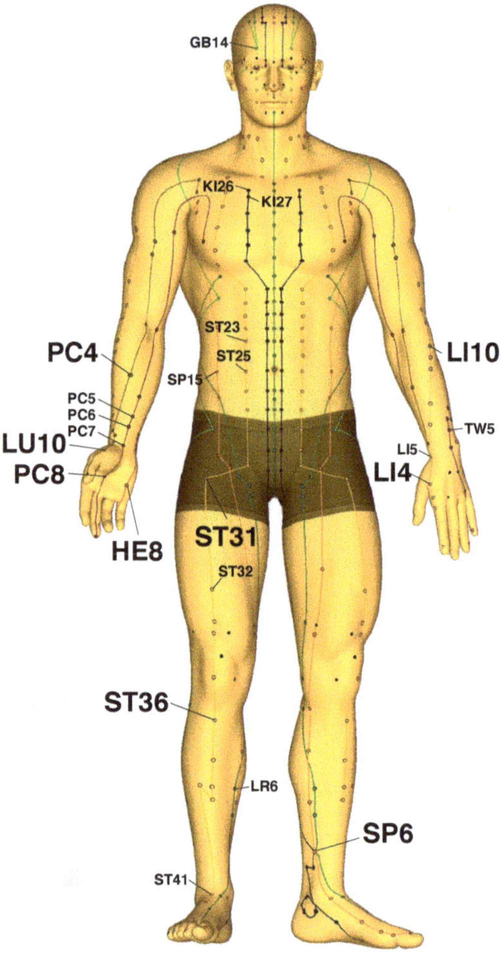

Protocol Points In-Depth
The following in-depth protocol points are a sum total of all the books I have studies and my personal experiences in treating clients.

Upper Body
1. Start at GB21 at the base of the neck. This releases tension in the body primarily in the connective tissues, reducing pain and opening the shoulders, head, and neck. As the highest point of the torso, it also may open people's ability to connect to their higher selves. Helping to connect the head and heart, this point allows the virtue of benevolence to be restored and renewed.
2. Working your way down to SI11 mid-shoulder blade opens the channels of the body, subduing pain and opening the shoulders and chest. Point name is "Heavenly Ancestor", which connects one to their ancestors to help guide them to a deeper wisdom and may also help to process dreams at a deeper level.
3. Next point is LI10 on upper forearm. This point helps relax and regulate stress in the body, alleviate pain, boost energy, and open the arms. Arm three-mile point, similar to the leg three-mile point, helps the arm and elbow loosen up and give more range of motion, revitalizing the arms, helping to feel nourished and receive the benefits of qi energy.
4. Next point is PC 4, five fingers from wrist crease. This point invigorates blood and removes blockages, cools and stops bleeding, and calms the spirit. It works well at removing heat from the blood, calming the heart, helping one get better sleep, and relieving many other mental and emotional disorders.
5. Next points are LI4 and LU10. Pressing both at the same time in the mid-corner of the webbing between the thumb and pointer finger, LI4, relaxes wild energy, strengthens the immune system, stops pain in the whole body, and helps with stress in the face and head. It empowers us to let go of what is no longer serving us, helping to drop the mask that covers up our true self and frees up the energy consumed by that mask. LU10 clears heat from the lungs and heals the throat. Both are the finishing point of the arms helping release and ground all wild energy of the

upper body, helping to bring warmth to the lungs to help burn away grief and can help to connect with our higher selves.
6. Next and last points are PC8 and HE8. Either press both at the same time or one at a time. PC8 is located between the second and third metacarpals halfway on the palm. PC8 clears heat from the heart and small intestine, revives consciousness, calms the spirit, harmonies the stomach, clears heat from the middle warmer, cools blood, regulates heart qi, activates the channel, and alleviates pain. It is the fire point of the pericardium channel in charge of helping remove heat and phlegm out of the heart to calm the spirit. HE8 is between fourth and fifth metacarpal halfway on the palm of the hand. HE8 helps calm the heart by quieting thoughts created by fright, fearfulness, sadness and worry.

Lower Body
1. Starting with BL54, the mid-point of the buttock, this point can open the channel from lower back to bottom of feet, helping to release pain in the lower back, hips, and legs. It helps connect the upper and lower body.
2. Working down to ST31, which is located in the crease of the upper thigh, named thigh gate. ST31 opens up the stomach channel to strengthen the qi/blood and opening the legs. When ST31 is used with ST36, it is a powerful combination to help relax the legs and give more uplifting energy to the body. It also may help a client sort and differentiate information easier making past traumas less challenging.
3. BL40 is just below the knee crease in the middle of the leg. BL40 has an amazing ability to open the legs, release back pain, unlock knees, and clear heat out of the blood and

extremities. It works to help create groundedness and a sense of stability in one's life.
4. ST36 just right of the shin bone and three fingers down from the outer knee in the muscle. It helps lift energy, calm the stomach, strengthen the immune system, calm wild energy, nourish the eyes, restore awareness, and relax the limbs. Leg three-mile point also helps to revitalize and resolve leg issues. It harmonizes qi and blood and also reestablishes upright qi, nourishing the muscles and empowers groundedness and centeredness. It helps a person to receive nourishment from food and life, allowing the ability to see more possibilities on one's life path.
5. Next, SP6 is three fingers up from inside ankle bone. It strengthens the stomach, calms stress, builds up kidneys, strengthens the legs and hips, helps urination, and increases circulation. Three yin junctions, the meeting point of spleen, liver, and kidney help to balance all three at the same time. SP6 also resolves dampness in the body and reestablishes upright qi helping with lethargy, depression, appetite, insomnia, infertility, and impotence. It has the great ability to empower us in many areas of our lives.
6. Next, BL60 on the outside of the ankle just aft of the ankle bone. This point cools heat, increases energy, calms wild energy, helps pain, relaxes connective tissue, and helps lower back problems. It helps balance yin and yang and nourishes the five elements of the body. It can also reestablish connection between heaven and earth—the foundation of life.
7. KD1 is on the bottom of the foot in the middle and just behind the front pad. This is the last point of the foot, and it is the only one on the bottom of our foot. This is a very important point to calm nerves, energize, lift up, clear heat, and ground the entire body. Out of all the points, this is the one that has the biggest effect on calming people. Called "bubbling spring of the body," it helps to connect to the earth and reestablish balance, both calming and revitalizing the mind and the body at the same time.

Chapter 4

Treatment Stories

A majority of my practice over the years has been working with veterans. PTSD and trauma of any kind however, may originate from different situations in which both men and women have experienced major trauma some time in their lives. The following client profiles demonstrate how shiatsu has helped people in the healing of their trauma.

Profile Client #1
Age 72
Gender – male
Combat experience - Vietnam war, Army (Tunnel Rat)

Signs and symptoms
Sleep issues: light sleeper, wakes at night, hard getting to sleep, vivid dreams
Nerve issues: jumpy, twitching muscles
Mental health issues: PTSD, paranoia, angers easily

Treatment details
PTSD/Trauma Protocol Points.
One-hour shiatsu session every week for six months.
Changed to once every month for next six months.
After a year of shiatsu treatments, he showed a marked decrease in anxiety and anger.
He experienced more days relaxed and happy and confident in his ability to manage his life.

History
This is one of my first and most memorable veteran clients with PTSD that I have treated. He served in the army as a tunnel rat. A tunnel rat in Vietnam was assigned the very dangerous job of being sent into the tunnels that ran across the jungles of Vietnam which were set with traps and armed Vietnam soldiers.

My client's main assignment for the Army was training other U.S. soldiers as tunnel rats. The soldiers were to go into these tunnels, seek out the Vietnam soldiers and kill them. The average life expectancy for this assignment was ½ hour.

Being under this constant stress for such a long time period has major effects on the nervous system, basically turning on his flight and fight response with no shutoff for years.

He had tried many ways to help with his severe trauma but with no success. When he tried acupuncture, his body would involuntarily move and pop out the needles.

The first time he came to me for a shiatsu session, within 15 minutes he was relaxed enough to fall asleep. At our next session, I asked about how he felt. He reported that after that first session, he had the best night sleep in a very long time. I was so inspired by this client that I began creating a specific shiatsu protocol for PTSD.

Profile Client #2
Age 45
Gender – male

Signs and symptoms
Sleep issues: Wakes up though out the night
Nerve issues: jumps and ducks around loud noises
Mental health issues: PTSD, people pleaser, anxiety, depression, hypervigilance

Treatment details
PTSD/Trauma protocol points.
One-hour shiatsu session every week for two months.
Changed to once a month for six months.
He showed vast improvements in only two months, sleeping though the night and an overall relaxed disposition.

History
He grew up on the south side of Chicago and was raised by his grandparents. He learned to survive on the streets with his winning personality - he could sweet talk anyone. Growing up on the streets, he saw a lot of violence, never knowing if he would be the one on the other end of the violence. If you talked to him he seem to have it all together but inside he was filled with heavy emotions. He had problems with work that also took heavy toll on him.

He when into a deep depression, not talking to anyone unless necessary. He stopped bathing and practicing personal hygiene. He came to me because he felt he had run out of options. With the support of a counselor and shiatsu treatments, he dramatically shifted the depression, sleep problems and felt relief from all his other symptoms.

He is still working though his issues but now has a strong footing and a better attitude to take on life with all its challenges.

Profile Client #3
Age 26
Gender – male
Combat experience –Syria and Iraq War, Marine infantry

Signs and symptoms
Sleep issues: Sleeps more than the normal 6/8 hours a night and is hard to wake up.
Nerve issues: Hyper alert, spaces out a great deal and felt ungrounded
Mental health issues: PTSD, daydreaming, detachment to life, unmotivated

Treatment details
PTSD protocol points.
One-hour shiatsu session every week for eight months.
Changed to one-hour shiatsu once a month for six months.

At first he took some time to respond but eight months into our shiatsu sessions he experienced a significant breakthrough with a new level of relaxation and calm alertness in his everyday life.

History

My client lost his right leg in a road boom (IED). Being the lead Hummer in his caravan made him hyper-aware of the danger of road booms. Two of his friends died in the explosion, and one friend was severely injured. My client suffered from both survivors' guilt and fear of traveling in vehicles.

It was challenging for him to even walk to the treatment center for our shiatsu appointments with his lost leg, but with great effort and determination, he never missed any of our sessions.

He showed a marked improvement in his anxiety four months into our shiatsu sessions and then began to plateau in his progress for a few months. Then he began to show a remarkable change; it was as if he came out of a long sleep. He was a new man super excited in life, filled with energy, and no longer experienced nightmares. Some of the symptoms came back but when they did, he was not so affected by them—he was a rock, unmoved.

Profile Client #4
Age 36
Gender – female

Signs and symptoms
Sleep issues: Wakes regularly at 3am and has trouble getting back to sleep.
Mental health issues: PTSD, anxiety, worry, always tired

Treatment details
PTSD protocol points
One-hour shiatsu session every week for two months
Changed to one-hour shiatsu once a month for six months

She responded fast once she got into a regular routine of shiatsu every week. These regular sessions helped her reach a new level of relaxation which allowed her to experience less stress in her everyday life.

History

My client had a history of major sexual trauma, so for her to experience shiatsu massage, was a major accomplishment for her right from the start. It took some time to build trust in our sessions and the more frequent the sessions, the more relaxed she became. She has been seeing a talk therapist for the past 5 years and from recommendation from her therapist, she began shiatsu treatments. With the combination of talk therapy and shiatsu's PTSD protocol, both helped her make significant progress in her recovery.

Rebuilding trust in people, especially in men has been a slow but rewarding process. She has a new level of trust in her world, allowing her to thrive rather than staying in survival mode on a daily basis. Her sleep improved after a month of treatment, getting the first full night sleep in years. She started to engage with more social groups, became more involved in her community and had a greater sense of being engaged in her world. She has been both an inspiration and confirmation that shiatsu can offer life-changing healing in one's life.

Profile Client #5
Age 74
Gender – Female
Combat experience - Vietnam war, Navy Nurse

Signs and symptoms
Sleep issues: vivid dreams, waking up thoughout the night
Mental health issues: PTSD, anxiety, depression

Treatment details
PTSD/Trauma Protocol points

One-hour shiatsu sessions every week for four months.
Changed to once every month for next six months.
She responded well to the treatments after a few sessions,
by relaxing more, eventually she began falling asleep during
our sessions.

History
She played a major role in saving lives working as a head nurse.
She was stationed on a medic ship of the shores of Vietnam for 4
years. She saw so much death and horrible injures come into her
station, it left a deep scar of pain and sadness in her. She is still
haunted by what she saw so many years ago showing that often
time does not always heal all wounds. Shiatsu gave her much
needed rest from the pain of continued depression and anxiety.
Shiatsu helped her relate to her therapist at a deeper level that
gained her a whole new depth to her recovery.

Chapter 5

Techniques Section

Pressure Techniques

- **Old Dog**
- **Hold**
- **Rocking shiatsu**
- **Cat shiatsu**
- **Four Finger shiatsu**

Old Dog (supportive pressure). Lean on the client with a light supportive pressure with the large parts of your body (such as the leg) to cover as much surface area as possible. The lean is equal pressure from both you and your client, giving them the reassurance of trust. This method helps the acupressure point therapy penetrate deeper.

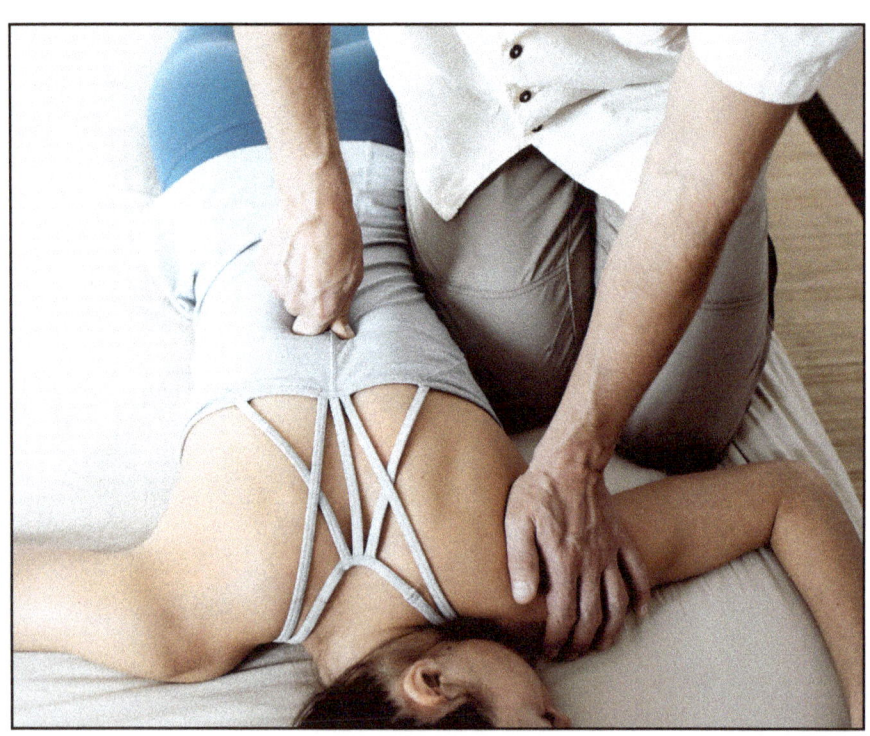

Hold. Stay on a pressure point for a set time to help release pressure. The average time is two to four seconds (if needed, stay for up to 30 seconds). This method helps your client relax more deeply by switching them into their parasympathetic nervous system, also called the "rest and digest mode".

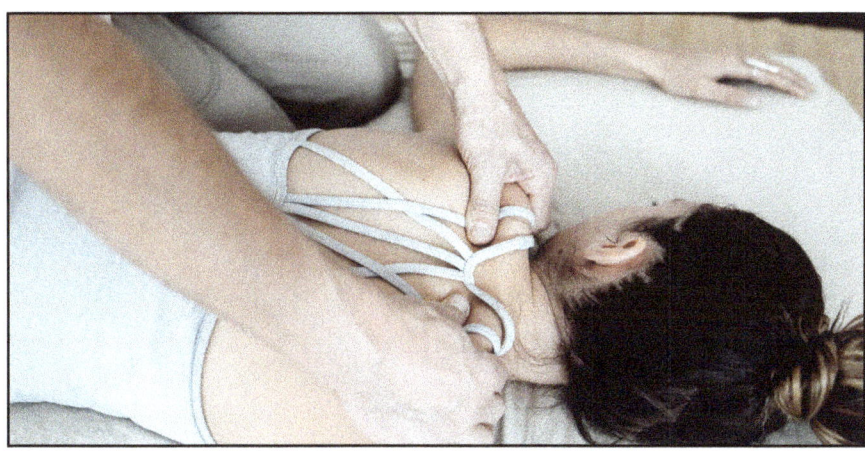

Rocking shiatsu. Apply pressure in the form of rocking back and forth continuously with your body. This method helps loosen the client's muscles as well as relax the nervous system.

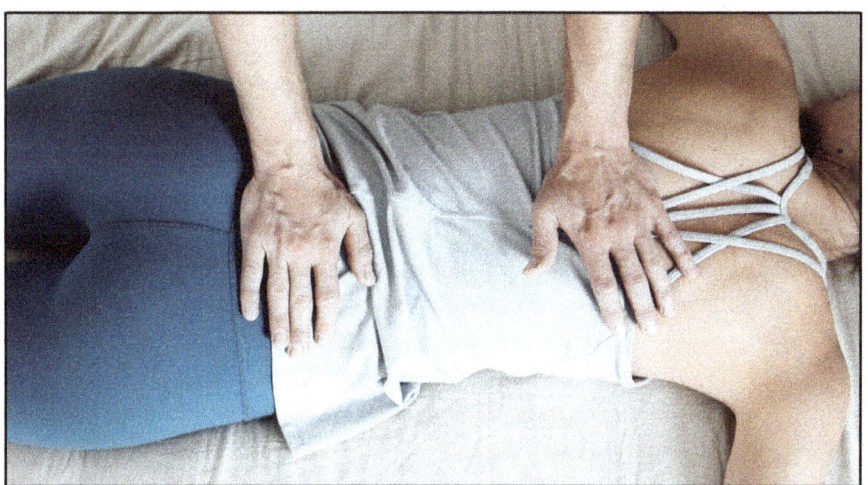

CAT shiatsu. In a kneeling position, push down with both hands, alternating pressure from one hand to the other.

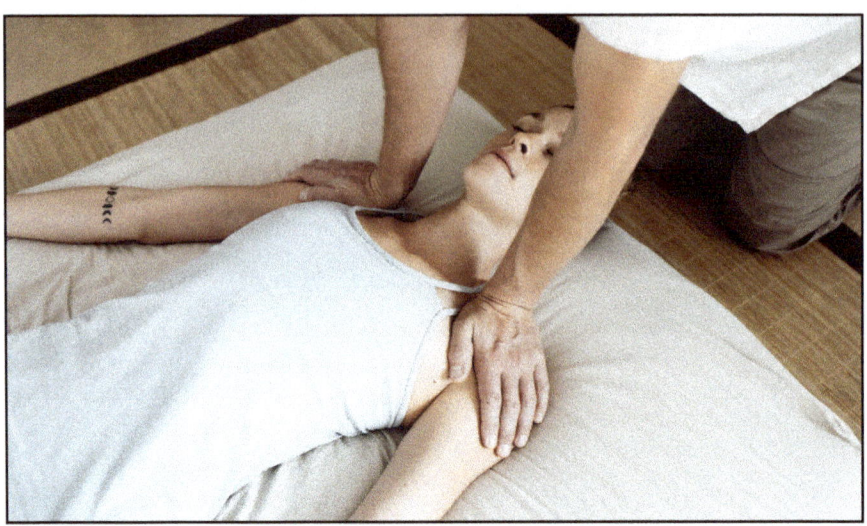

Four fingers. Use the pads of all four fingers on one hand to diagnose the meridians and the Hara to treat imbalances.

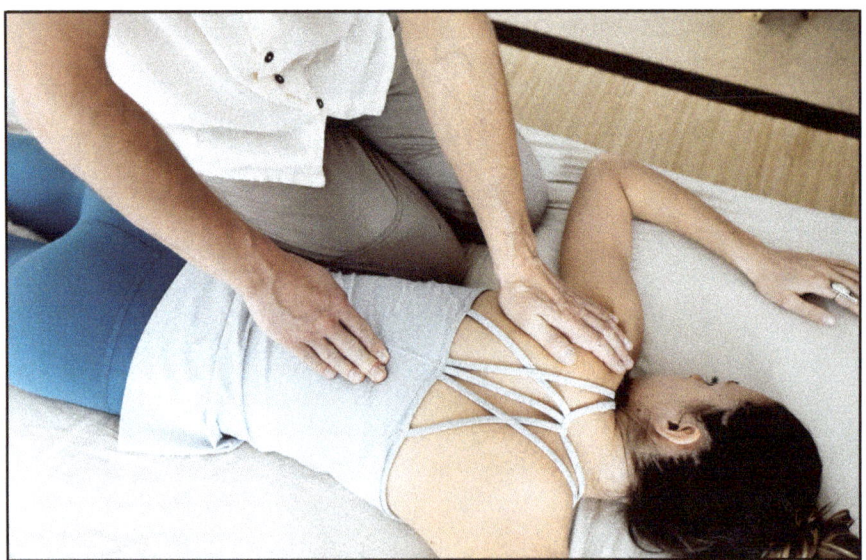

Hand Techniques

- **Brush off**
- **Heel or blade of the hand shiatsu**
- **Mother hand**
- **Palm shiatsu**
- **Rotation shiatsu**
- **Squeeze and twist shiatsu**
- **Stretching shiatsu**
- **Thumb shiatsu**
- **Four fingers shiatsu**
- **Thumb and index knuckle shiatsu**

Brush off. Finishing a sequence by lightly brushing down the area that was just massaged. This technique is used when ending your work or when transitioning into the next sequence.

Heel or blade of the hand shiatsu. Use stronger and more contracted pressure to treat target areas.

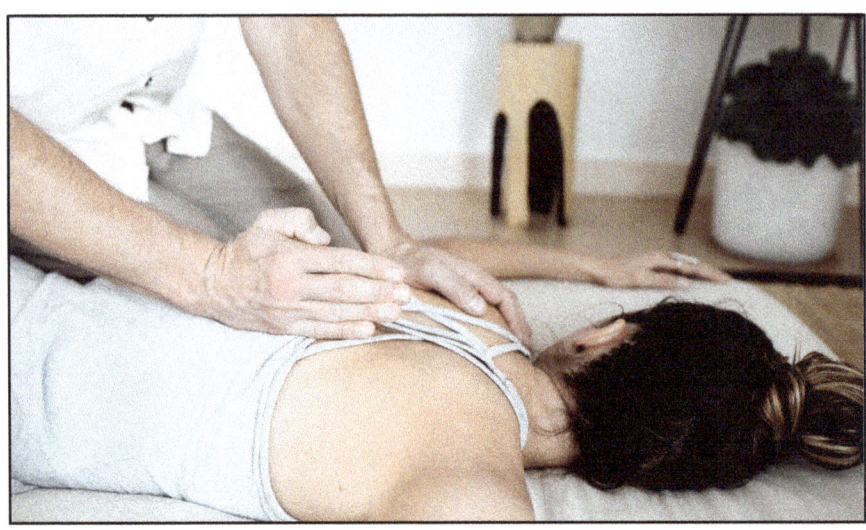

Mother hand. Hold one hand stationary to help calm the nervous system and to sense the reaction of the body of the effects of pressure from the other hand.

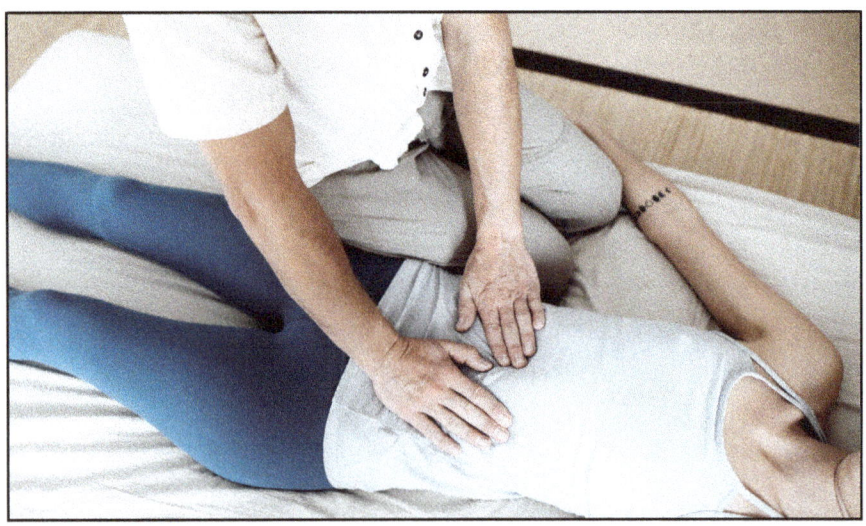

Palm shiatsu. Use the whole palm to give soft, larger area pressure. This technique is used to understand what is going on in the body.

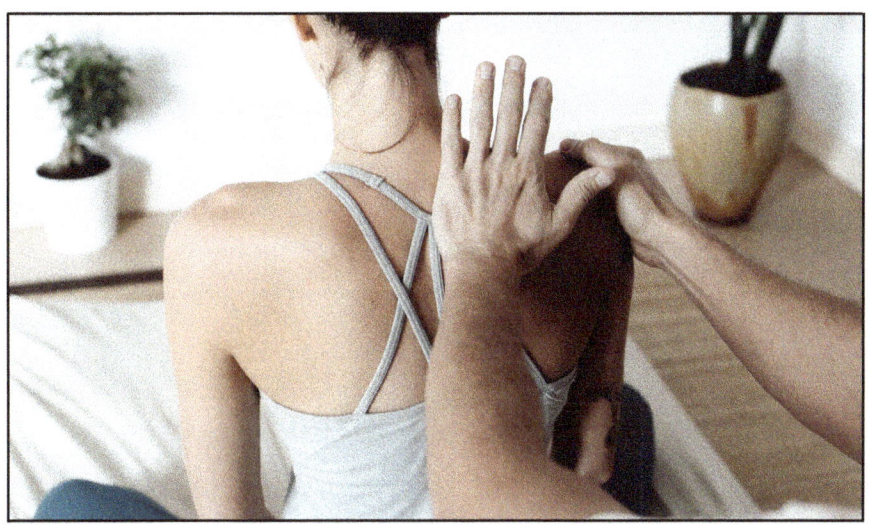

Rotation shiatsu. Rotating the arm at the shoulder, elbow, and wrist; rotating the leg, knee, and foot.

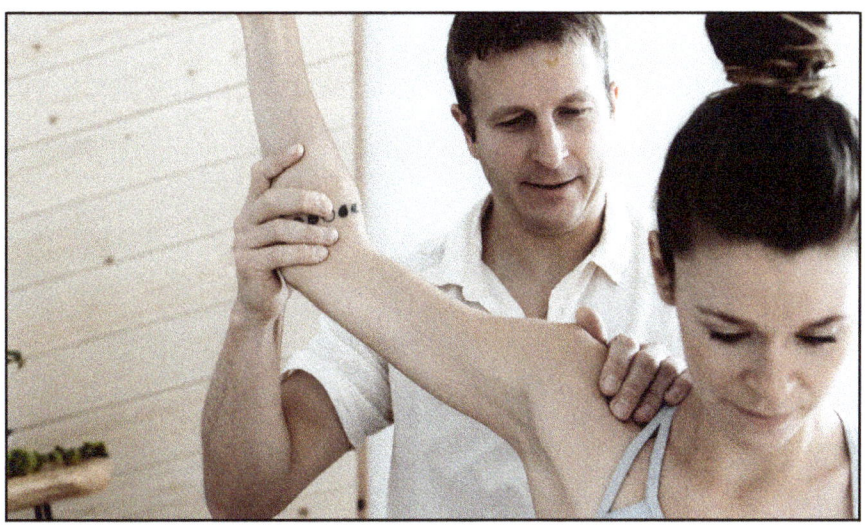

Squeeze and twist shiatsu. Alternate pressure between both hands with one hand staying in place while the other hand moves and presses down on a specific body part.

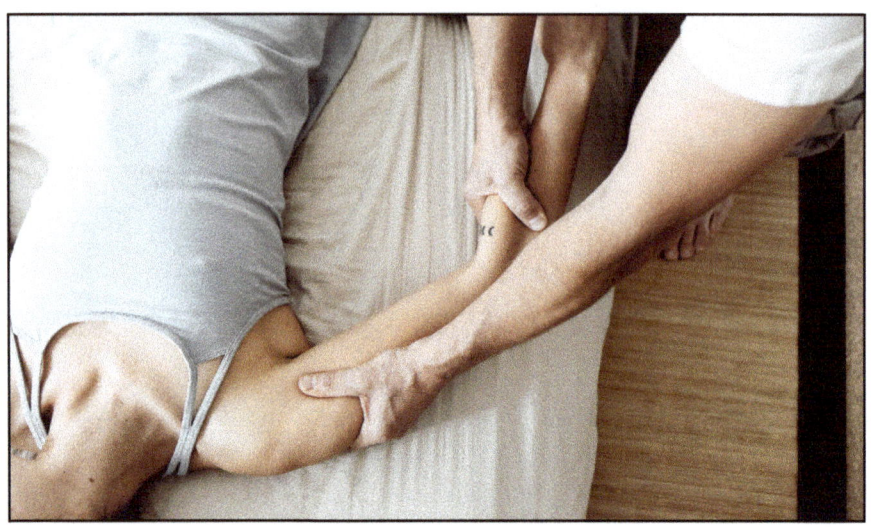

Stretching shiatsu. Extend a body part to the full range of motion using slow and steady pressure.

Thumb shiatsu. Squeeze the thumb and hand together to provide pressure on a point.

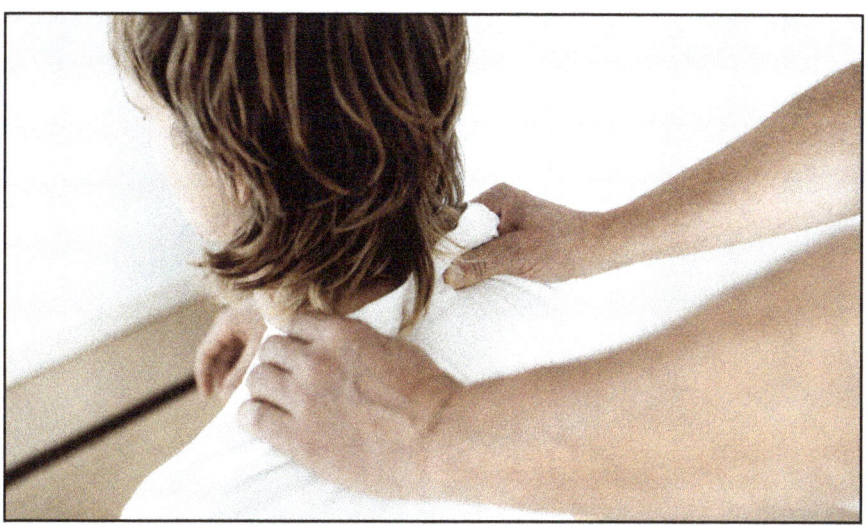

Thumb and index knuckle shiatsu. Both photos on this page show how the thumbs are supported against the bent index finger enabling you to press with both fingers at the same time.

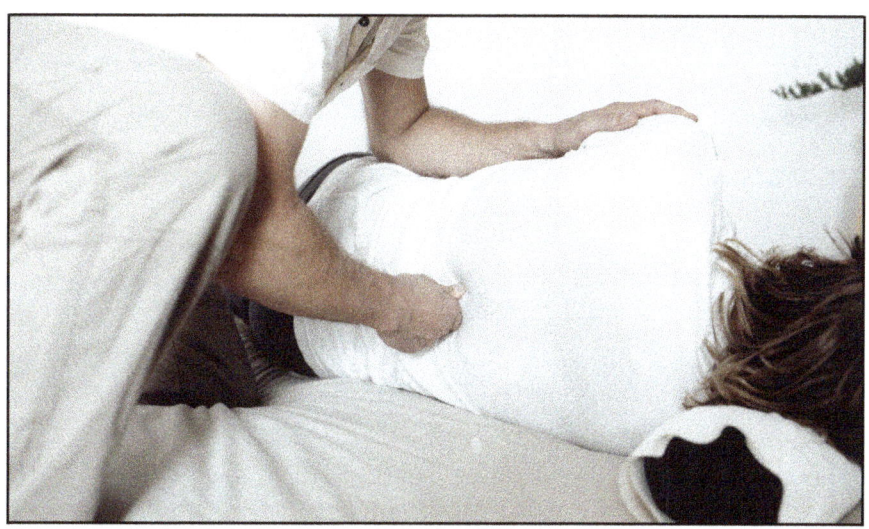

Leg/Arm Techniques

- **Elbow shiatsu**
- **Forearm shiatsu**
- **Knee shiatsu**

Elbow shiatsu. Use the point of the elbow for more direct pressure or the Ulna for more widespread pressure.

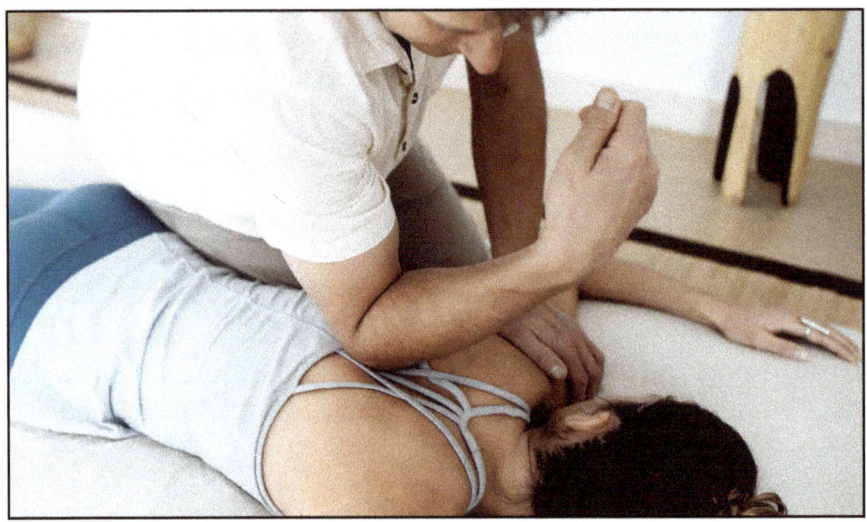

Forearm shiatsu. Similar to elbow shiatsu; use the blade of your forearm to press down to create an even pressure.

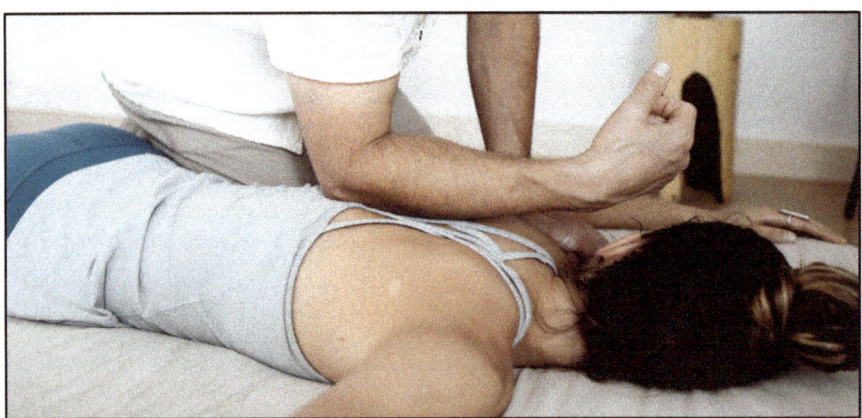

Knee shiatsu. Use the area just below the knee with the support of your arms on the other side of the client's body to control the pressure.

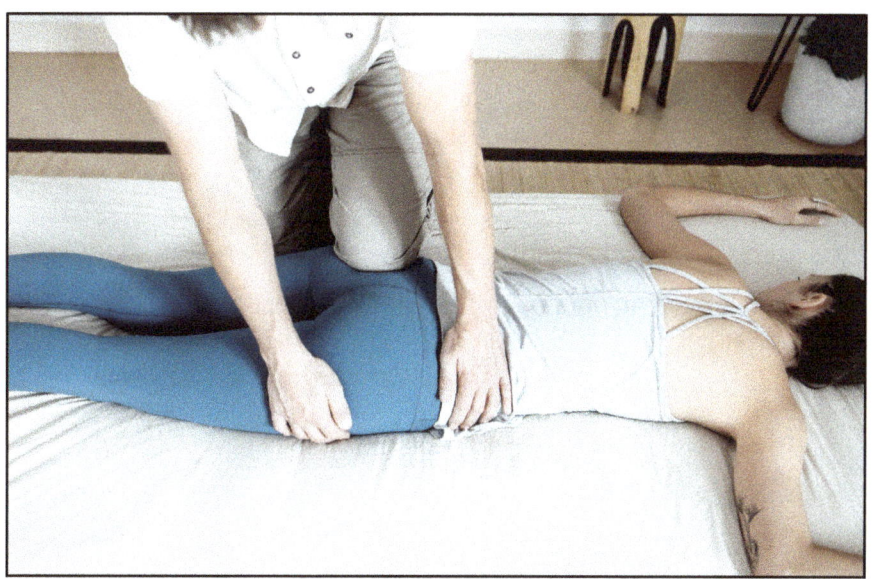

Chapter 6

Short Sequence (Chair)

The short sequence is done in the seated position. Note: If you cannot reach your client's back, remove this part from the sequence.

This sequence includes palpation of entire meridians along with the following set points:
(Refer to meridian chart Appendix B for point reference)

BL-10; GB-9, 12, 14, 20, 21; GV-21, 24; LU-1
(neck and head)
BL-13, 14, 15, 17, 18, 20, 23, SI-11 (back)
GB-21 (shoulders)
HE-5, 7, 8; LI-4, 5, 10; LU 10 ; SI-7; PC-4, 5, 6, 7, 8 (arms)
Red = Protocol points

Thumb shiatsu shoulders: hold both hands on top of the shoulders with a gentle squeeze between your thumb and fingers. Repeat next shiatsu on both sides of the shoulder.

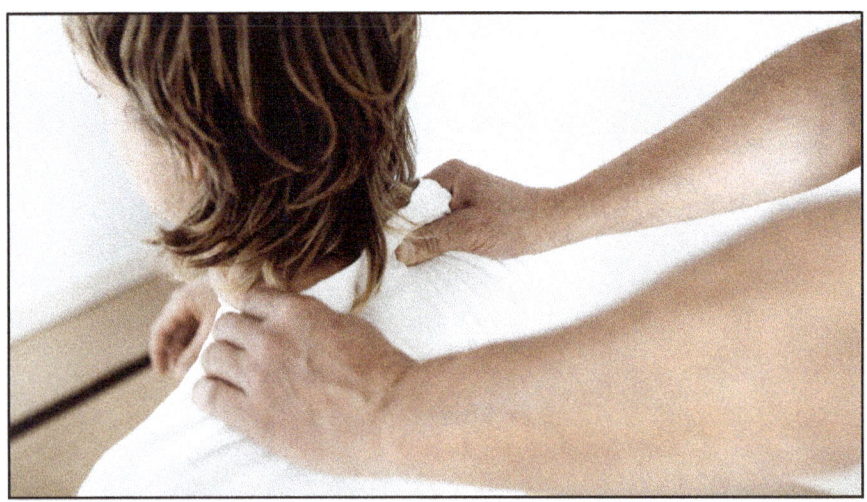

Palm shiatsu down back: holding the shoulder with one hand, with the other hand, gently palpate down one side of the spine.

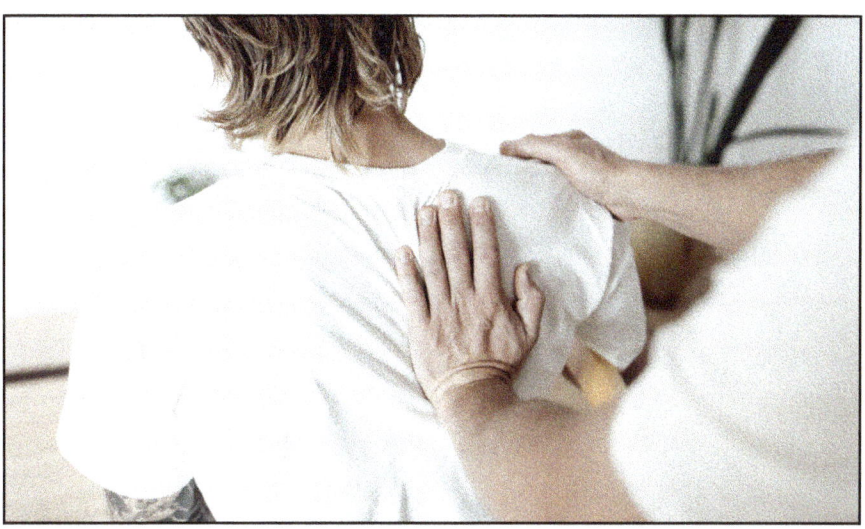

Forearm shiatsu down back: keeping one hand on the shoulder and using the blade of the forearm, press down in three places; the upper back next to spine, middle between spine and shoulder blade, and next to shoulder blade.

Forearm shiatsu top shoulders: press down with the blade of your forearm on the trap (trapezius muscle at the base of the neck).

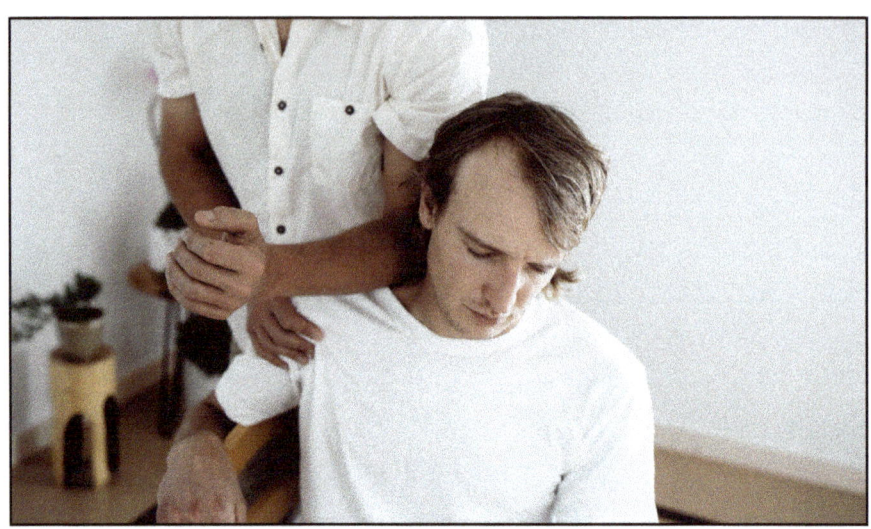

Elbow shiatsu chest: elbow shiatsu upper chest.

Squeeze and twist shiatsu arm: squeeze and twist the arm with both hands and as you move down the arm twisting in opposite directions.

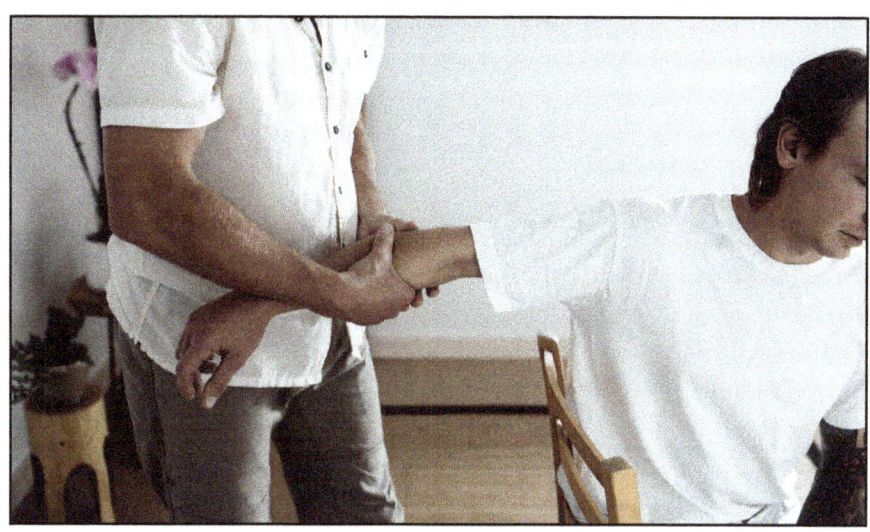

Thumb shiatsu hand: shiatsu both sides of the hand.

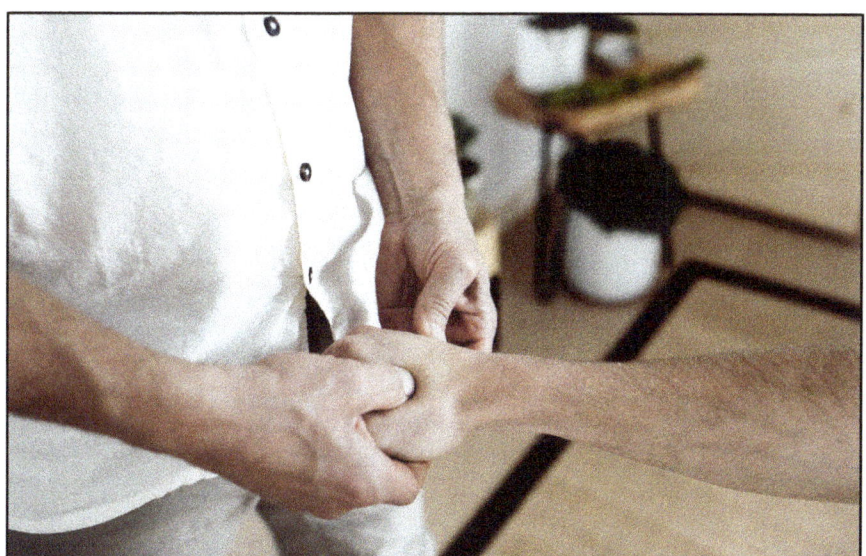

Rotation shiatsu of arm: holding the arm by the elbow, rotate in full from the shoulder socket.

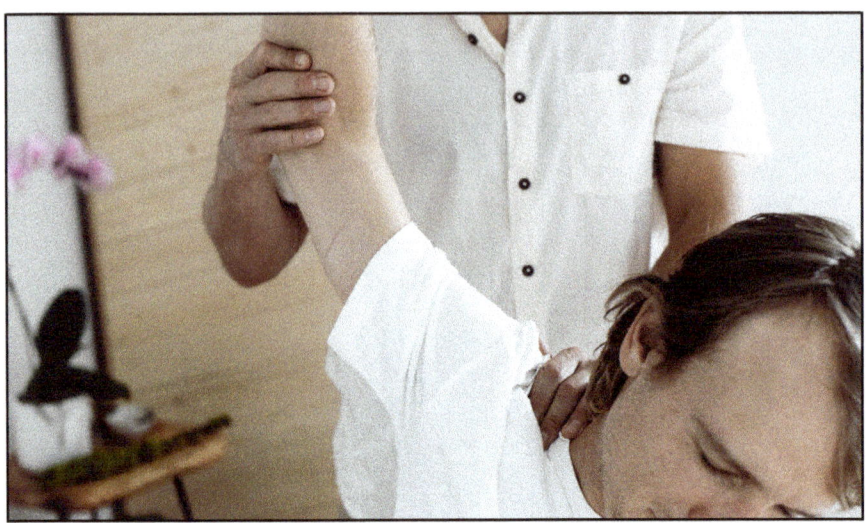

Stretching shiatsu arm overhead: pushing the arm you just performed rotation shiatsu on, press your arm on client's head and slightly behind it, and at the same time with the other hand, push into the stretching arm holding just below the neck on the shoulder.

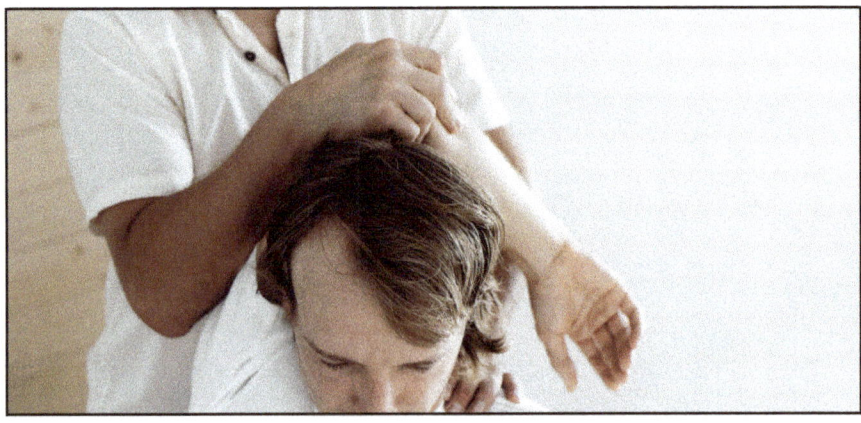

Repeat the past shiatsu on both sides of the body.

Thumb shiatsu shoulders: hold both hands on top of shoulders with a gentle squeeze between your thumb and fingers.

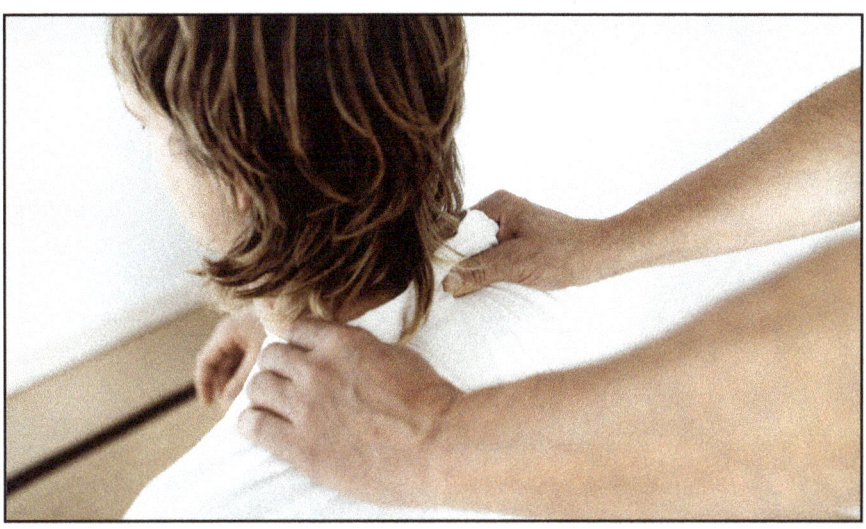

Thumb shiatsu neck: hold the head at the forehead with one hand using your same arm's elbow on the client's shoulder, with the other hand perform shiatsu the neck.

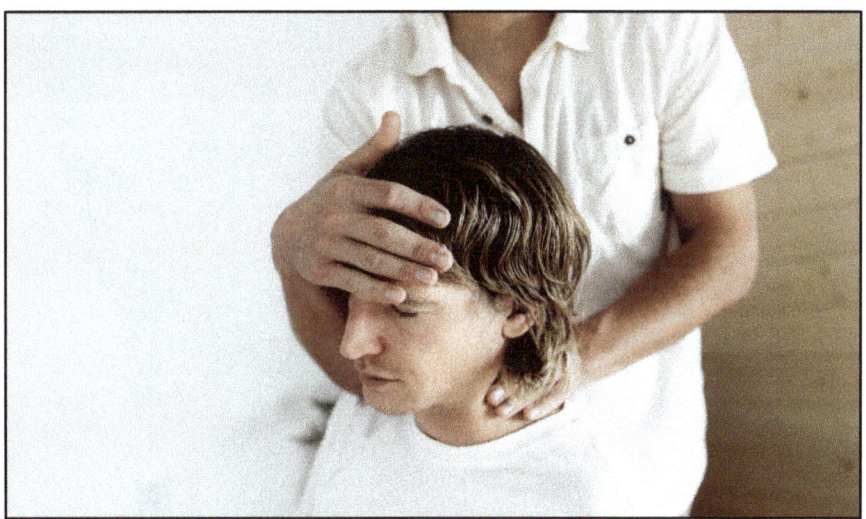

Repeat thumb shiatsu on both sides of the neck.

Thumb shiatsu shoulders: hold both hands on top of shoulders with a gentle squeeze between your thumb and fingers.

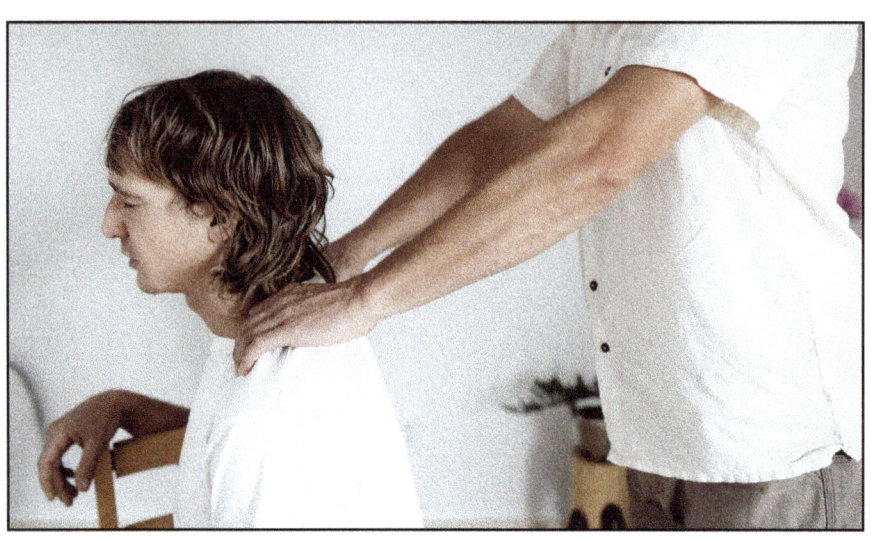

Sweep off, brush off shoulders and back to finish the session.

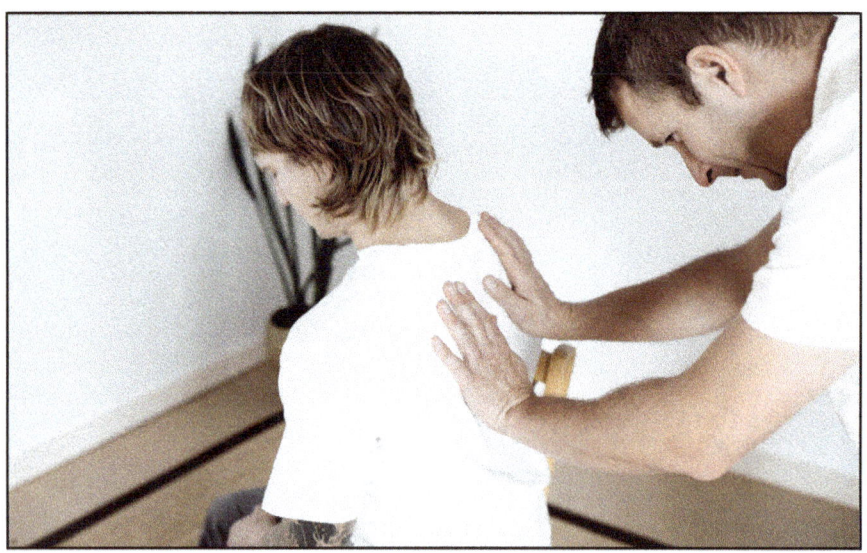

Chapter 7

Long Sequence (Mat)

The long sequence includes three different positions: seated, side lying, and face up (supine).
Note: If you cannot reach the back, remove that part from the sequence.

Seated (Mat)

This sequence includes palpation of entire meridians along with the following set points:
(Refer to meridian chart Appendix B for point reference)

BL-10; GB-9, 12, 14, 20, 21; GV-21, 24; LU-1
(neck and head)
BL-13, 14, 15, 17, 18, 20, 23, SI-11 (back)
GB-21 (shoulders)
HE-5, 7, 8; LI-4, 5, 10; LU 10 ; SI-7; PC-4, 5, 6, 7, 8 (arms)
Red = Protocol points

Thumb shiatsu shoulders: hold both hands on top of shoulders with a gentle squeeze between thumb and fingers.

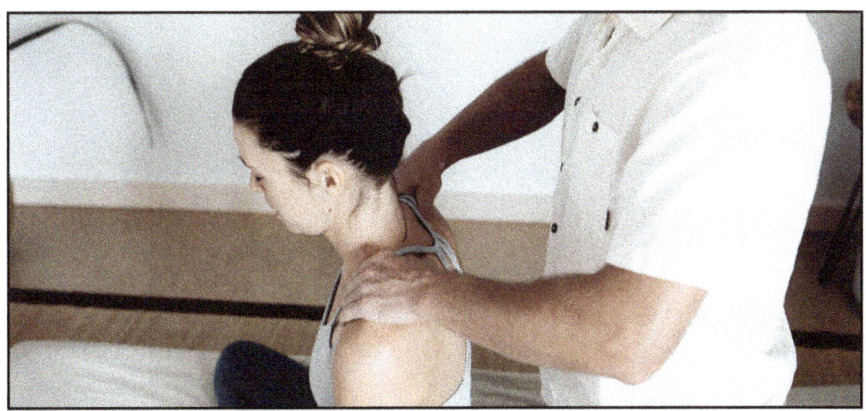

Palm shiatsu down back: holding the shoulder with one hand, use your other hand to gently palpate down one side of the spine.

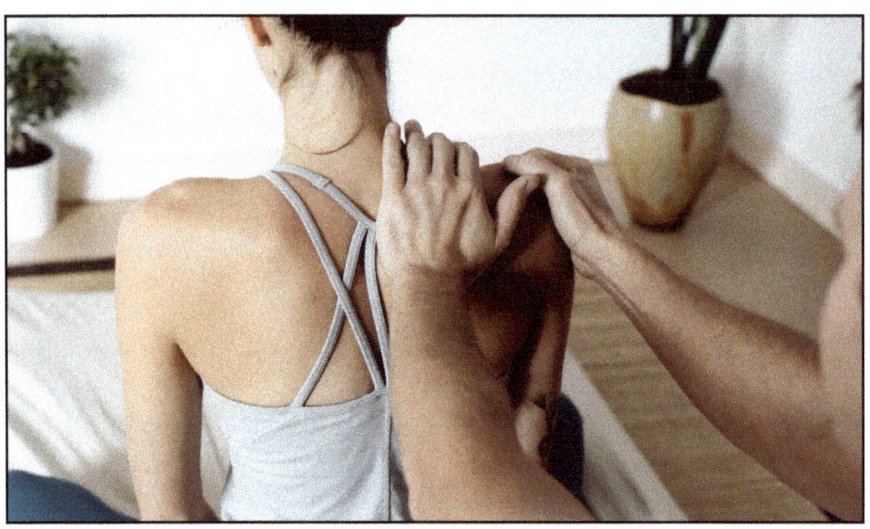

Forearm shiatsu down back: keep one hand on the shoulder and using the blade of the forearm, press down the upper back, next to the spine, middle between the spine and shoulder blade, and next to the shoulder blade.

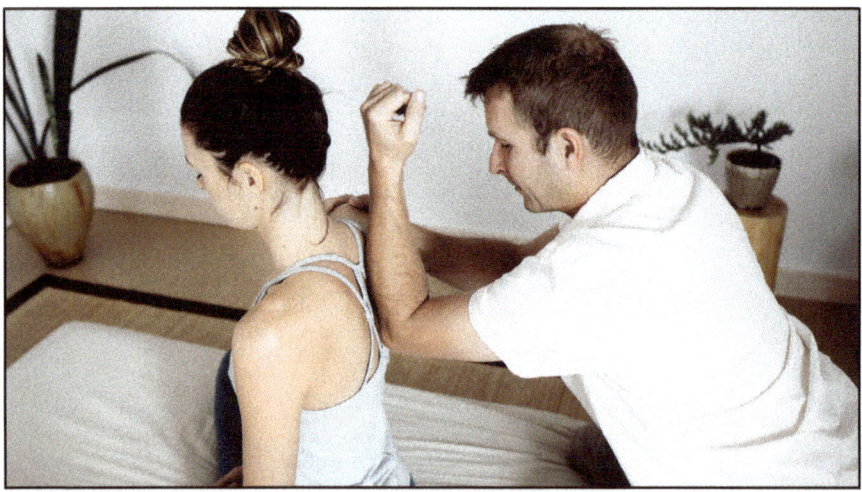

Forearm shiatsu top shoulders: pressing down with the blade of your forearm or hand blade on the trap.

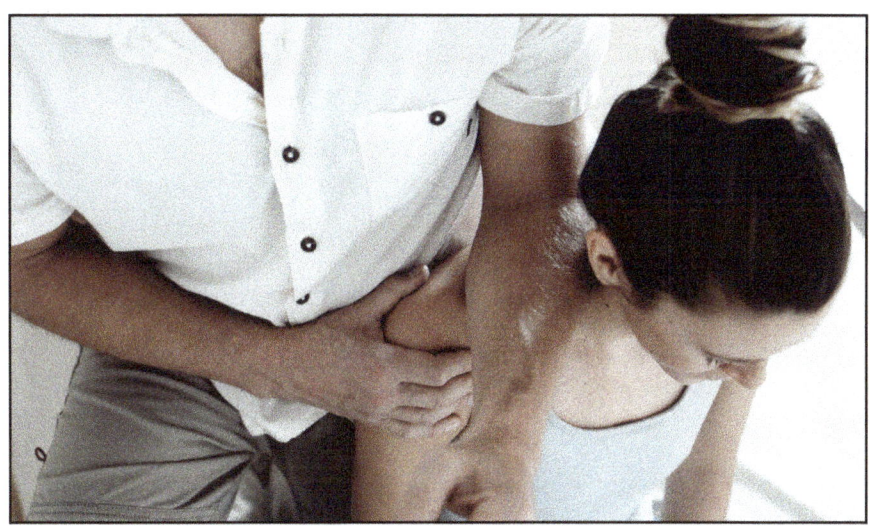

Elbow Shiatsu chest: elbow shiatsu upper chest.

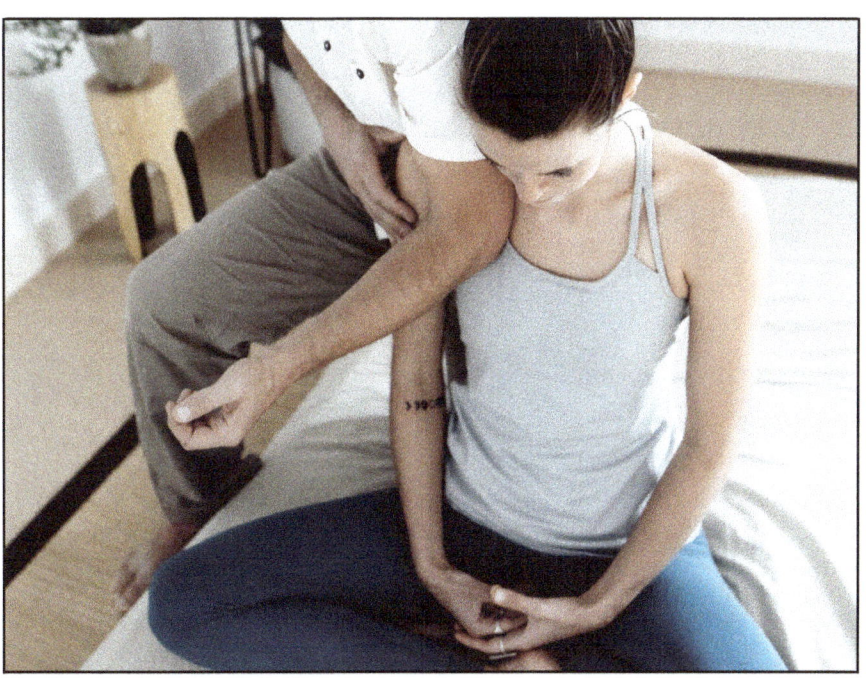

Forearm shiatsu arm: forearm shiatsu down whole arm from shoulder to hand.

Stretching shiatsu hand: start by stretching from the tips of fingers until the whole hand is stretched back.

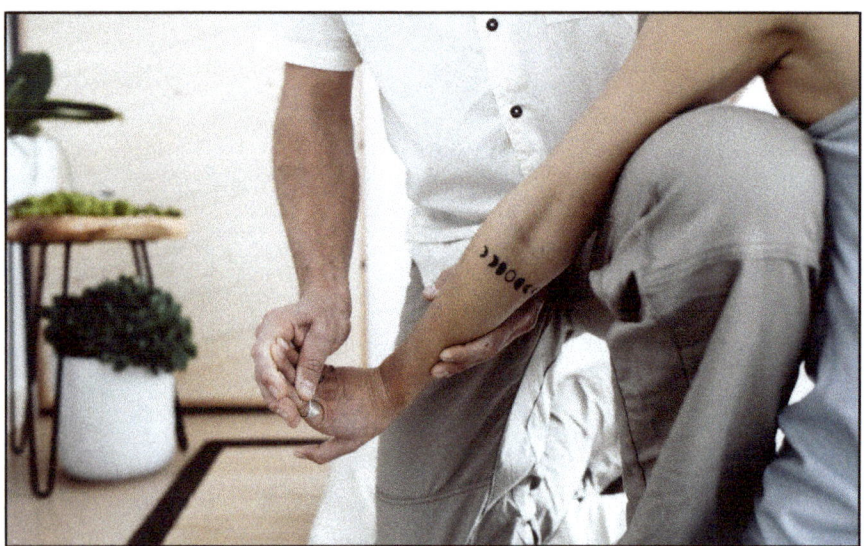

Squeeze and twist shiatsu arm: squeeze and twist the arm with both hands, and as you move down the arm twist in opposite directions.

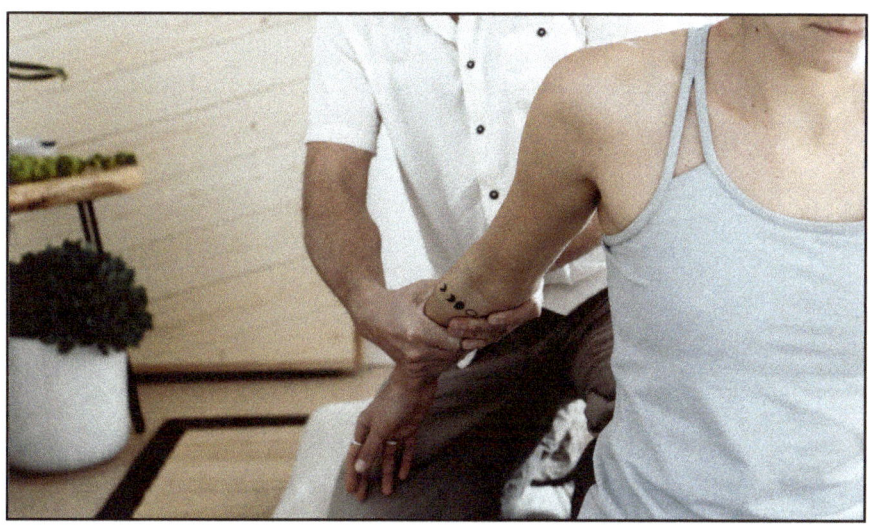

Rotation shiatsu of arm: holding the arm by the elbow, rotate the full range of the arm socket.

Stretching shiatsu arm overhead: pushing the arm you just performed rotation shiatsu on, press it to the head and slightly behind it and at the same time with other hand, push the opposing direction of the arm, holding just below the neck on the shoulder.

Repeat the past sequence on both sides of the body.

Thumb shiatsu shoulders: hold both hands on top of shoulders with a gentle squeeze between thumb and fingers.

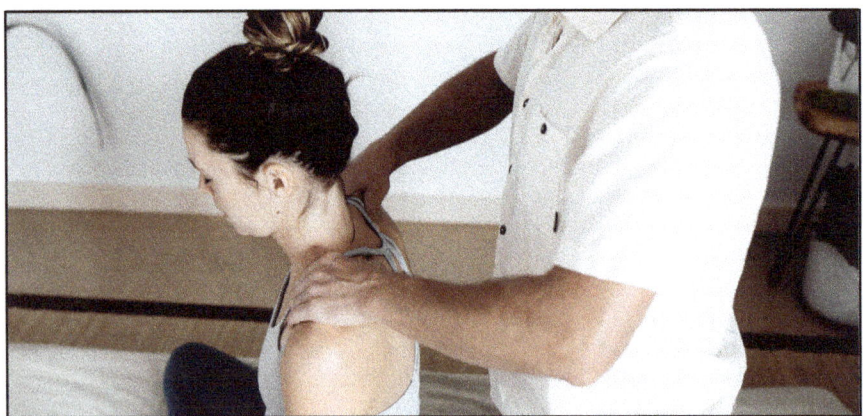

Thumb shiatsu neck: hold the head at the forehead with one hand and with the other hand massage the neck on one side with the thumb.

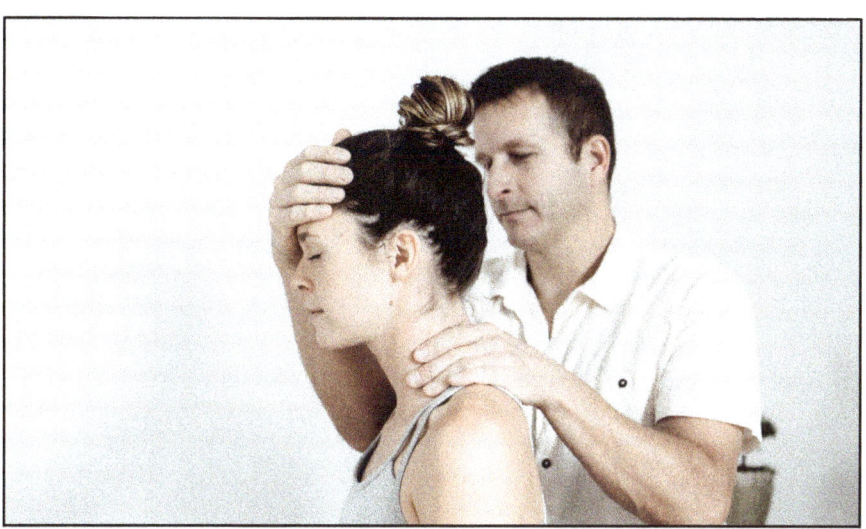

Repeat thumb shiatsu on both sides of the neck.

Thumb shiatsu shoulders: hold both hands on top of shoulders with a gentle squeeze between thumb and fingers.

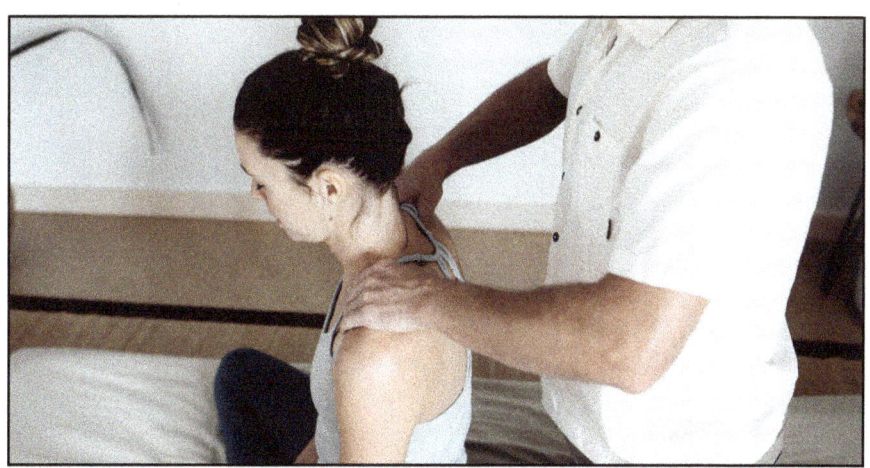

Brush off shoulders and back to finish the session.

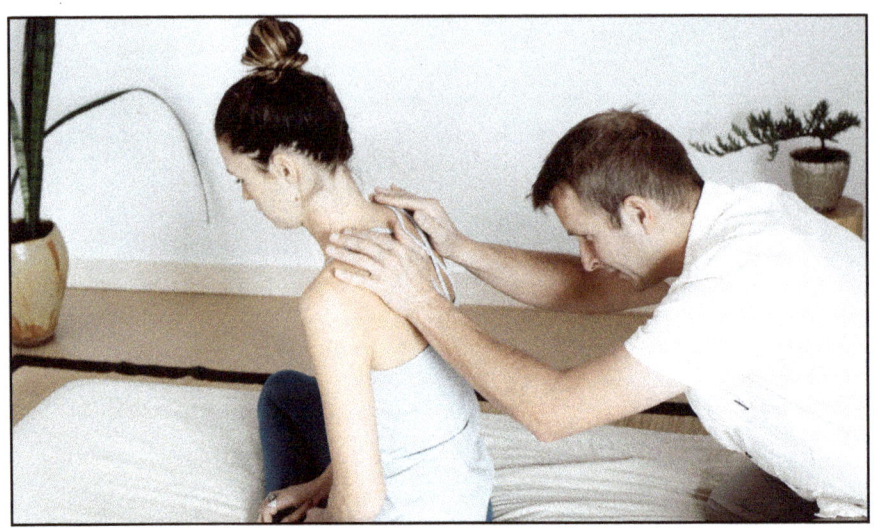

Side Lying (Mat)

This sequence includes palpation of entire meridians along with the following set points:
(Refer to meridian chart Appendix B for point reference)

BL-10; GB-9, 12, 14, 20, 21; GV-21, 24; LU-1
(neck and head)
BL-13, 14, 15, 17, 18, 20, 23, 27, 28, 29, 30, 31, 32, 33, 34, 38, 54, 61; SI-10,
11 (back and glutes)
HE-5, 7, 8; LI-4, 5, 10; LU-10; P-5, 6, 7, 8; SI-7; TW-5
(arms)
KD-1, 6, 9; LV-6; ST-31, 36, SP-6, BL-40, 60 (legs)
Red = Protocol points

Palm shiatsu hip: press gently over the sacrum, move up over the gluteus and illicit crest, with Mother hand on top of the hip.

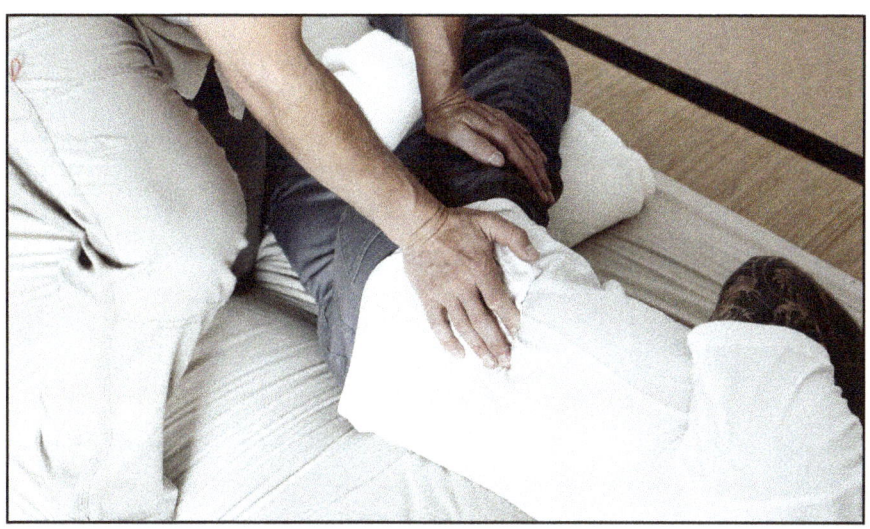

Palm shiatsu thigh: press down the thigh along the IT band.

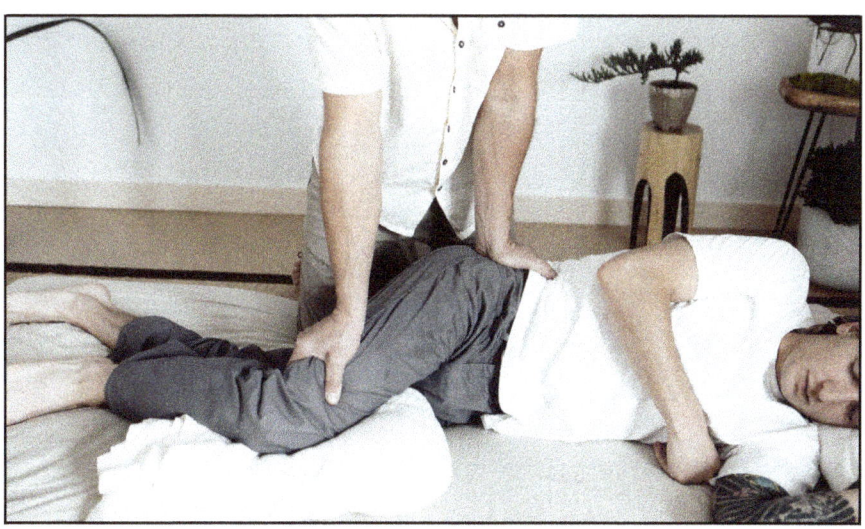

Thumb shiatsu calf: press with both hands and behind the shin bone from knee to foot.

Palm shiatsu foot: shiatsu top of the foot.

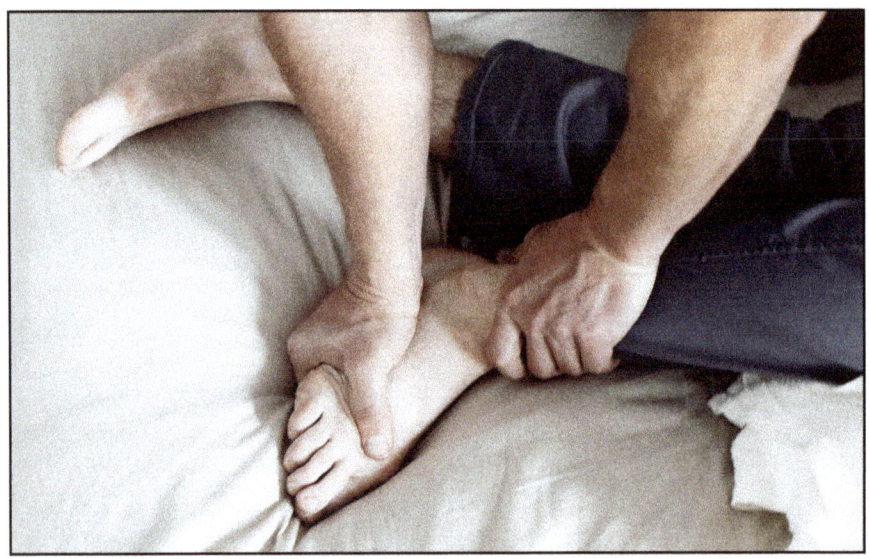

Knee shiatsu inside leg: pushing down with both hands, one hand on the hip and other hand on the foot, use your knee to softly press down the inner thigh from hip to knee.

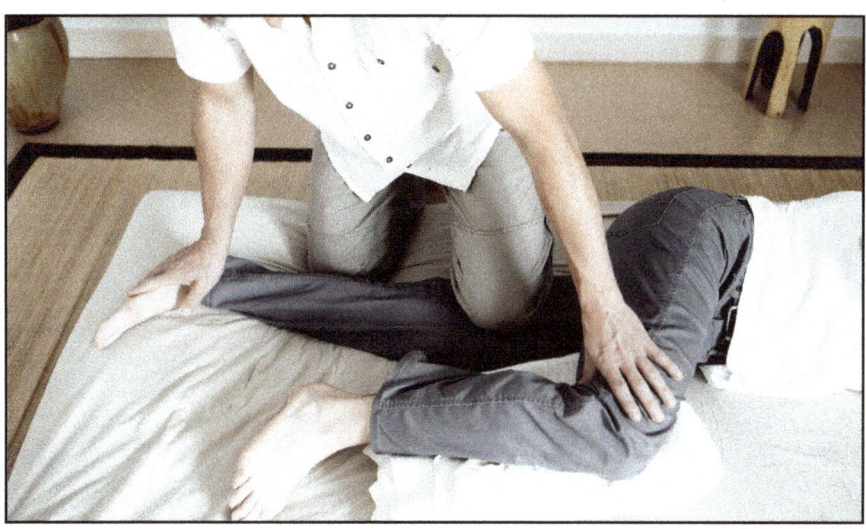

Knee shiatsu foot: massage the bottom of foot with your knee.

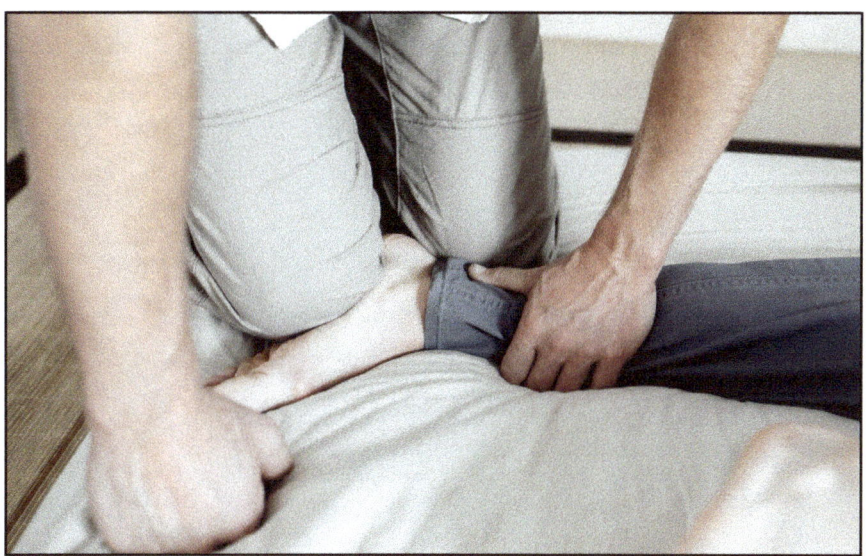

Palm shiatsu head: move to the head, pressing down with both hands on the side of head, one hand remains still and the other hand moves down and back up the head.

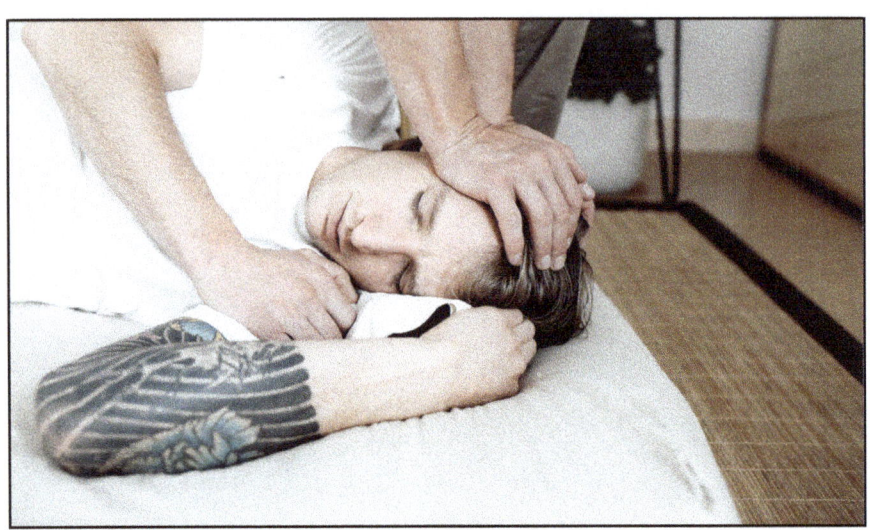

Thumb shiatsu neck: hold shoulder with one hand and use the thumb with the other hand to gently massage the neck.

Elbow shiatsu chest: using your elbow, press in and down along the upper part of the chest.

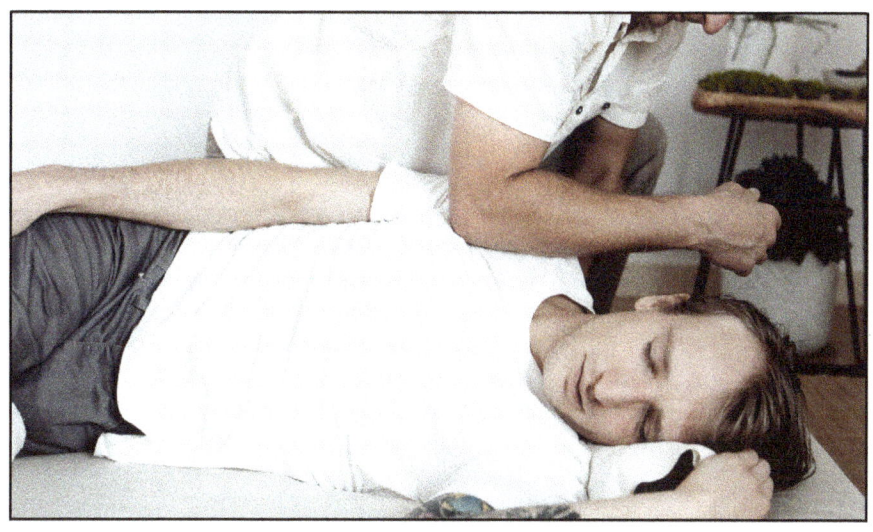

Forearm shiatsu arm: push down with your forearm blade on the upper arm and hold for 1–2 seconds.

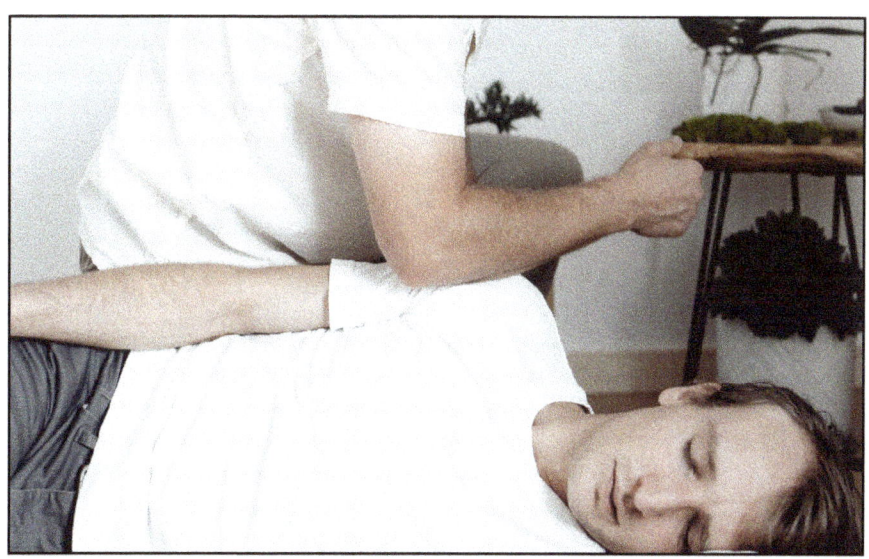

Thumb shiatsu neck: holding the front of the head with one hand, use your other hand to massage down both sides of the spine in the neck.

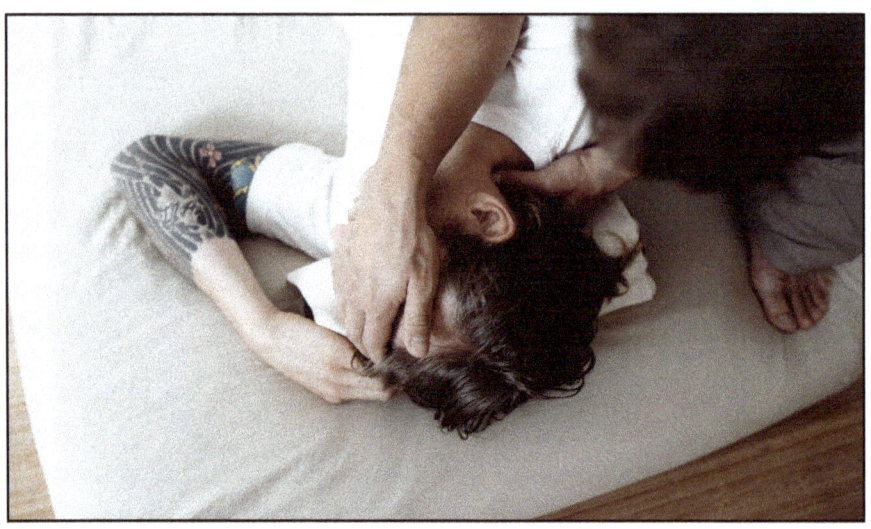

Thumb and index knuckle shiatsu spin: move your Mother hand to the shoulder and with the other hand continue down the spine to the sacrum.

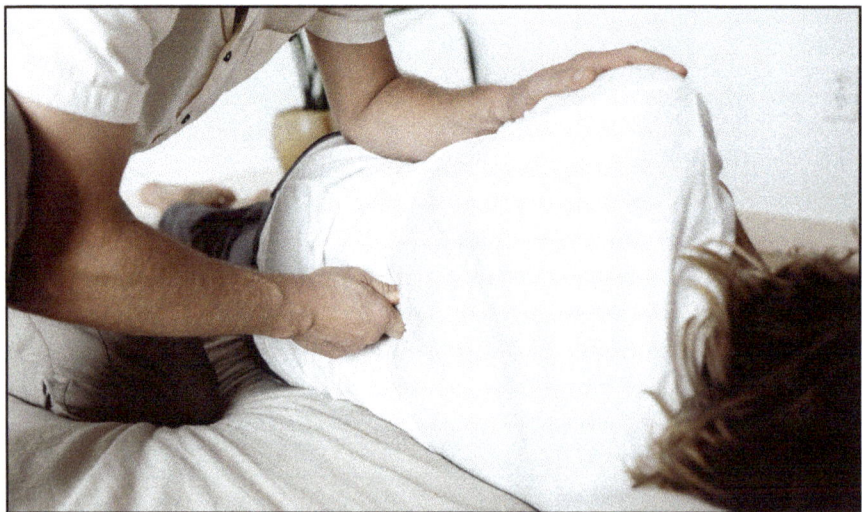

Elbow shiatsu trap: massage the top of trap with elbow, holding Mother hand on the back.

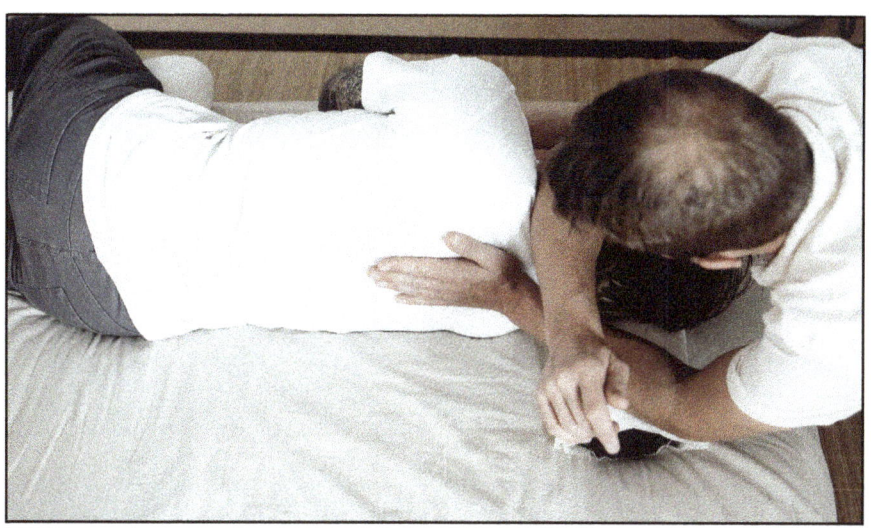

Elbow shiatsu back: shiatsu the upper part of the back above spine from neck to hips.

Squeeze and twist shiatsu arm: massage the arm by squeezing and twisting from the shoulder to the hand.

Rotate shiatsu of arm: rotate the arm around the full range of the shoulder socket.

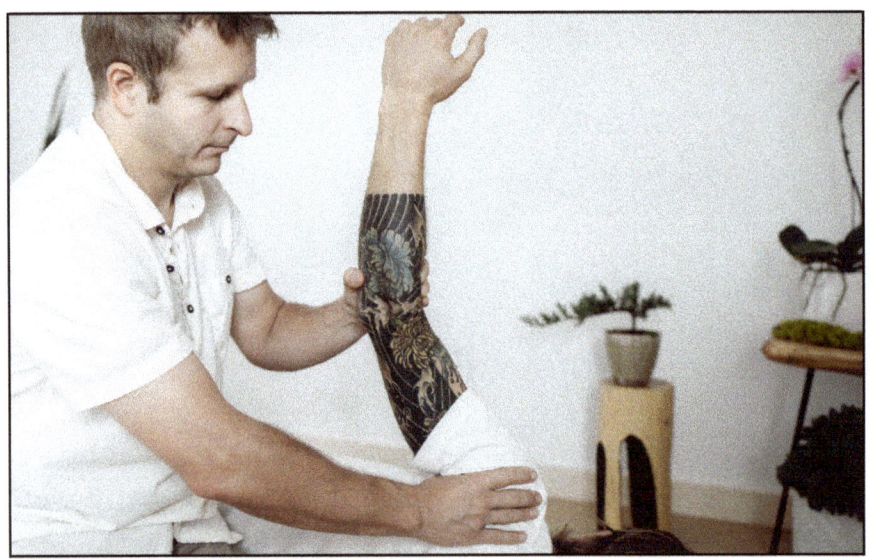

Stretching shiatsu arm overhead: push the arm from the elbow, and with your other hand, press the hip, pushing both in opposite directions for a side stretch.

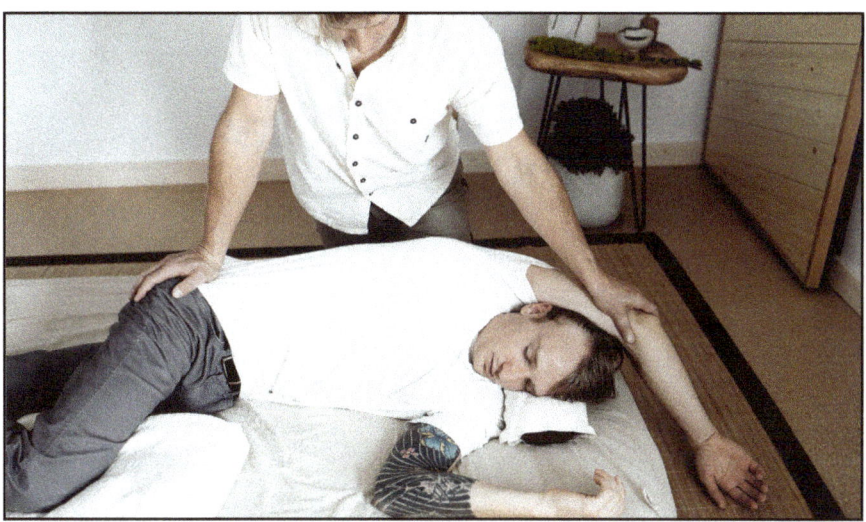

Forearm shiatsu side of body: with your body next to your client's, use your forearm blade to press the side of the body, working your way from shoulder to hip.

Knee shiatsu back of leg: holding the hip with one hand, and with the other hand hold the knee, pull both hands at the same time, pushing your knee into the back of the leg in opposite directions from hip to knee.

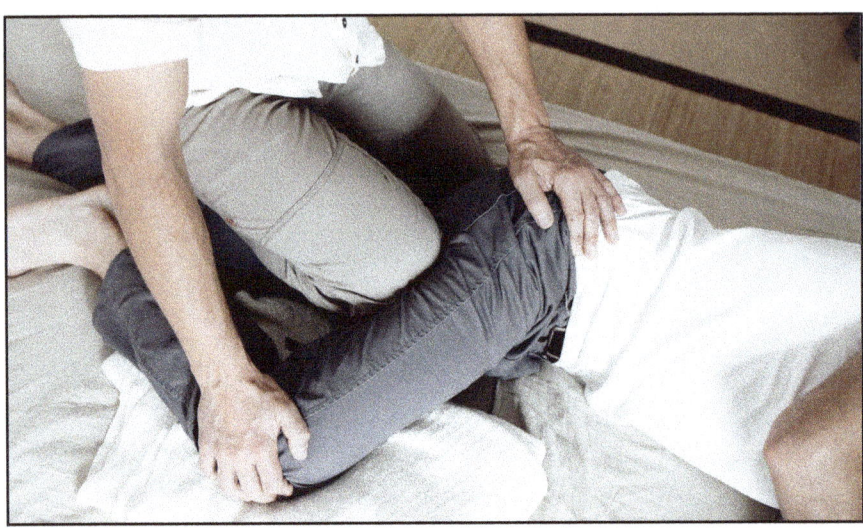

Thumb shiatsu shin: with both hands, press on the muscle next to the shin bone from knee to ankle.

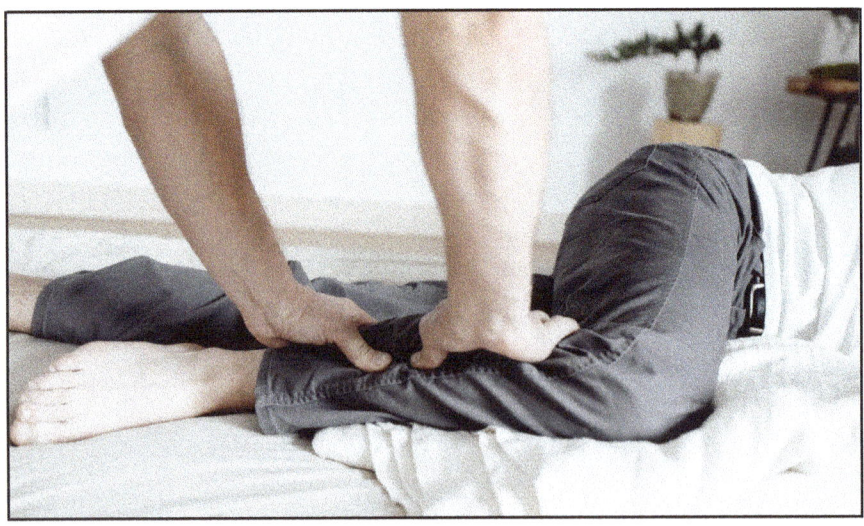

Palm shiatsu foot: massage the top of the foot.

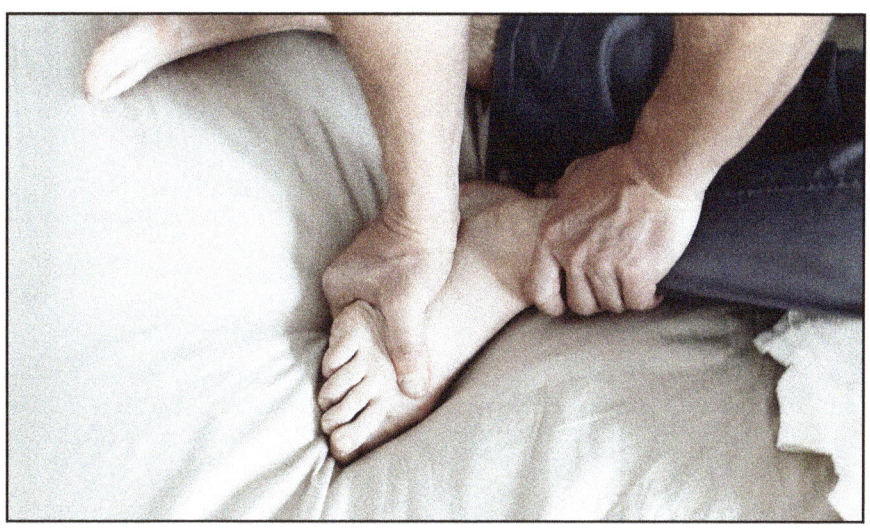

Knee shiatsu inside leg: pushing down on your hands, one on the hip and the other on the foot, softly press down the inner thigh from hip to knee.

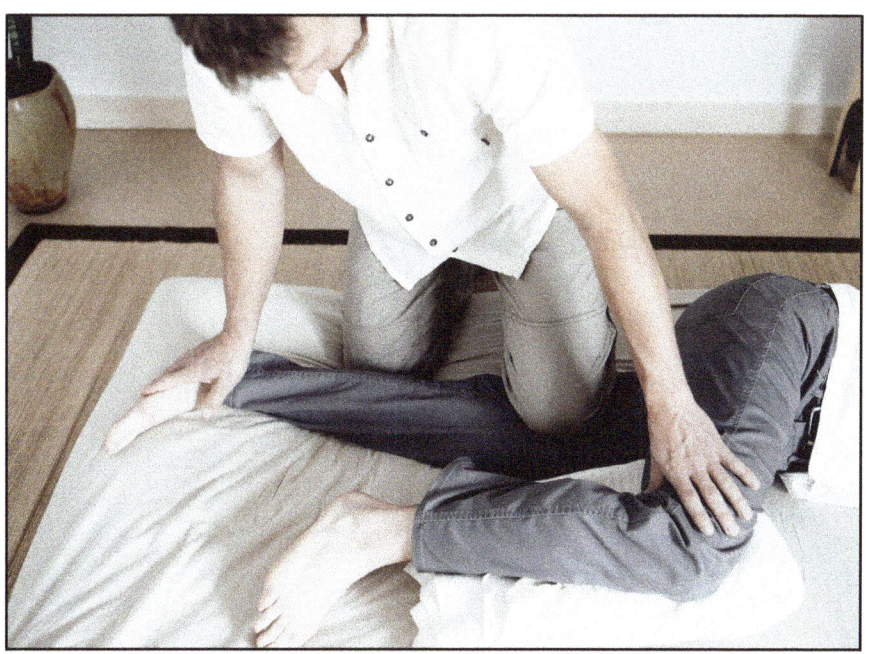

Knee shiatsu foot: massage the bottom of foot with elbow or hands.

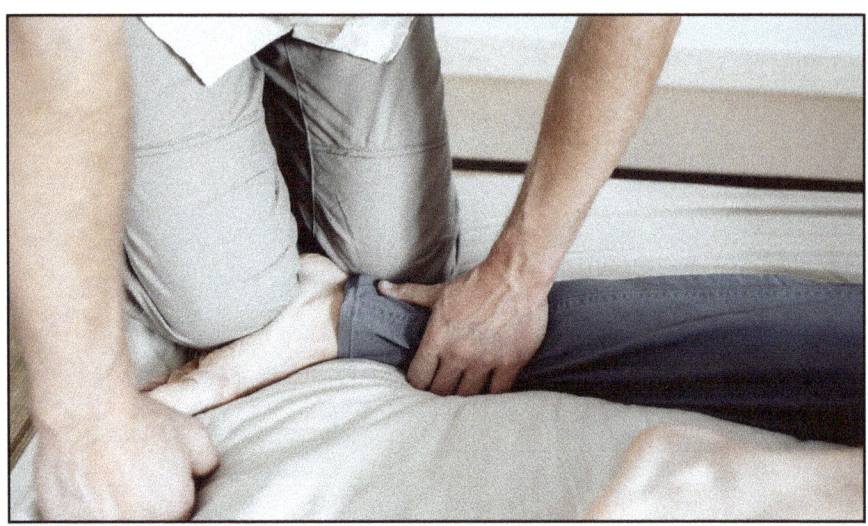

Rotating shoulder: holding the shoulder with both hands, one on either side, rotate the whole shoulder, then push the shoulder in the opposite direction of the bent leg, giving a spinal twist.

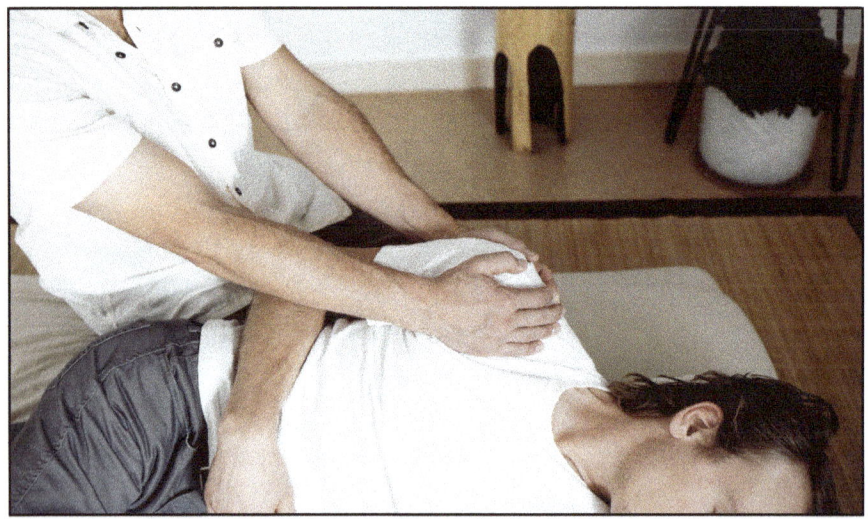

Stretching shiatsu leg: pull the top leg back just above the knee and at the same time, push your knee in the top of the hip bone, pulling the leg back.

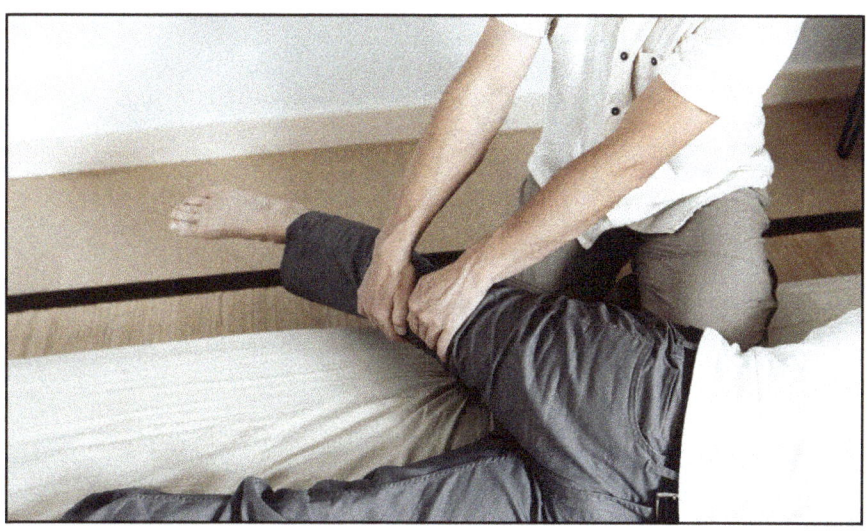

Brush off: brush off from the shoulders down the back.

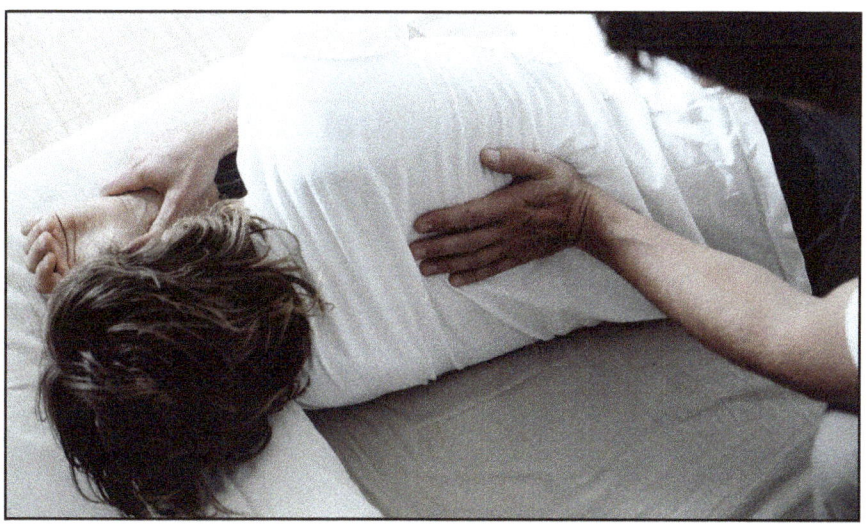

Repeat the entire sequence on the other side of the body.

Face-up (Mat)

This sequence includes palpation of entire meridians along with the following set points:
(Refer to meridian chart Appendix B for point reference).

CV-6; K-26; SP-15; ST-23 (Hara)
ST-31, 36, SP-6, BL-60, KD-1 (legs)
LI-4,10, PC-4, 8, HE-8, LU-10 (arm)
Red = Protocol points

Palm shiatsu stomach: place both hands on abdomen, one will be the Mother hand and the other will palpate all the organ systems related to the Hara.

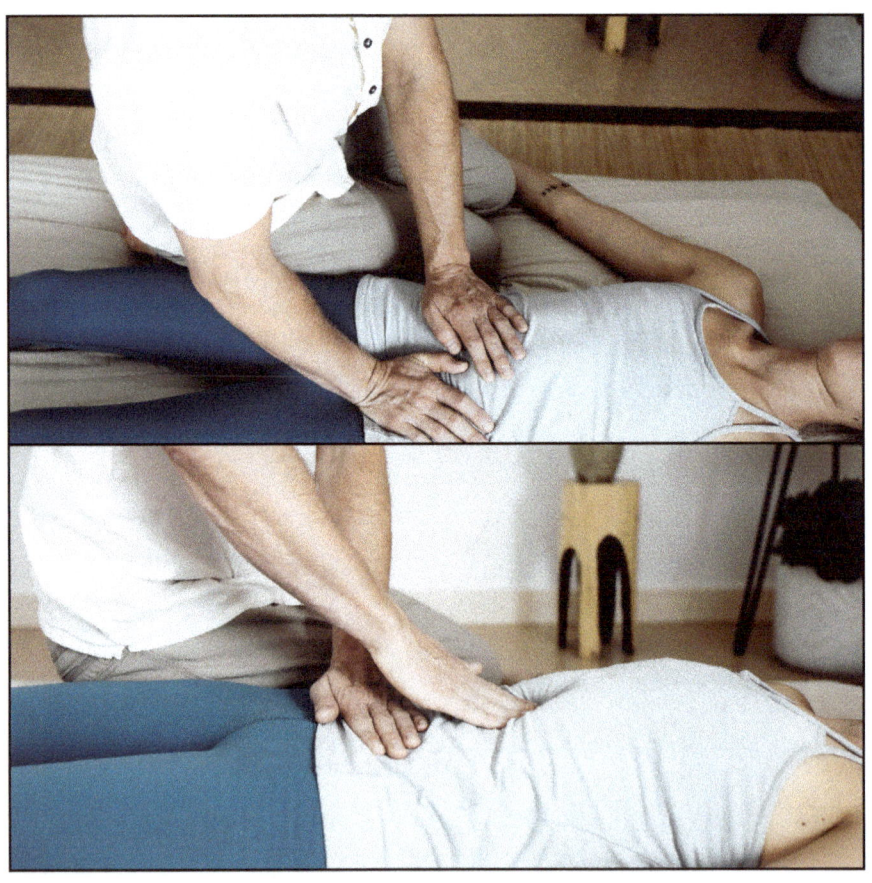

Palm shiatsu legs: press on the illogic crest, then work your way down both legs front and side of the legs, ending at the feet.

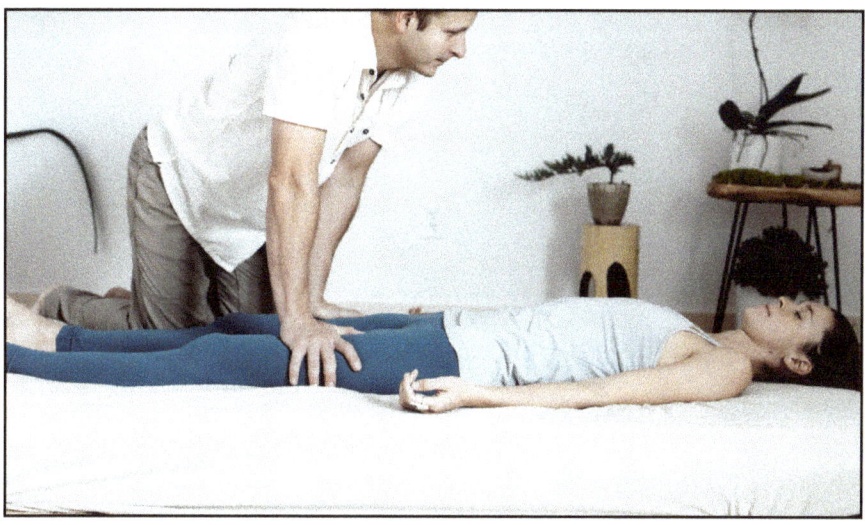

Rotate shiatsu feet: press the feet in and out then rotate them in both directions.

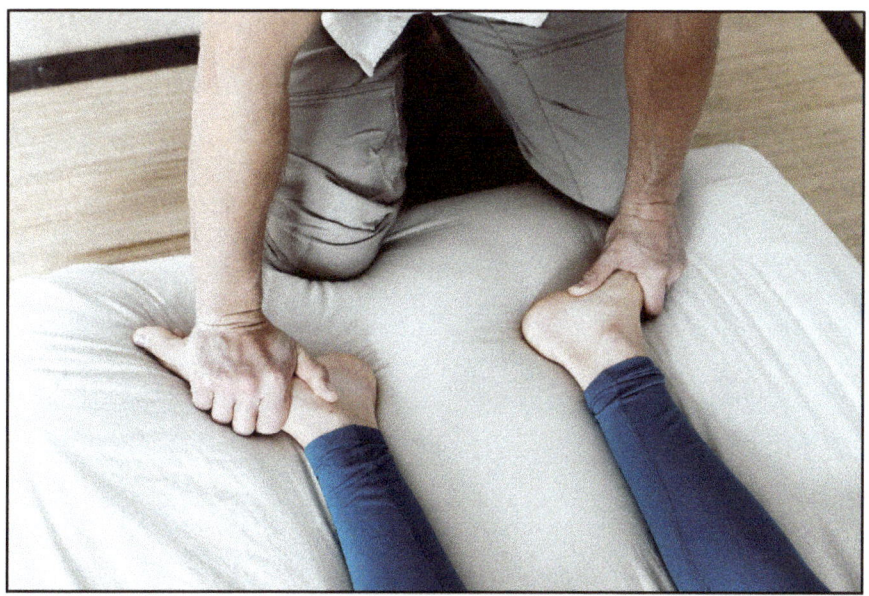

Squeeze and twist shiatsu arms: move up the arms squeezing and twisting from the shoulder to the hands.

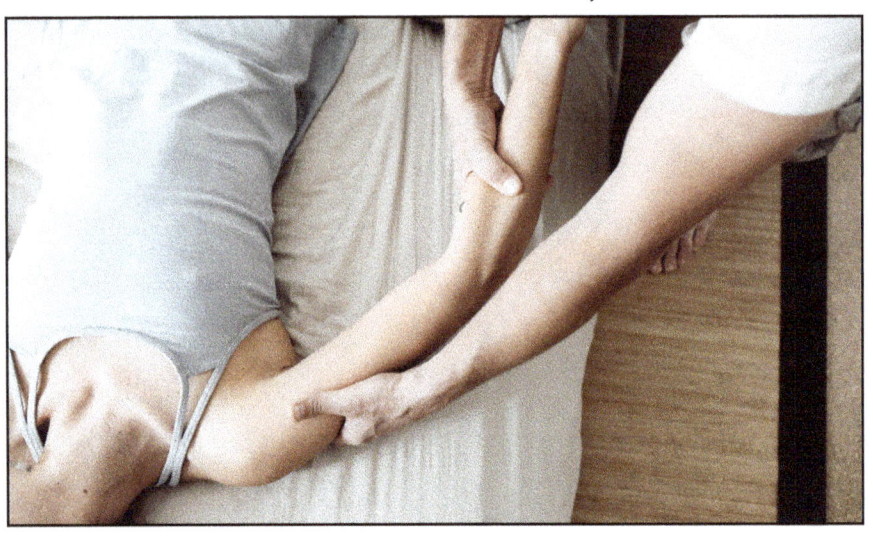

Rotate shiatsu arms: hold the arm at the elbow and rotate the whole arm in full range of motion.

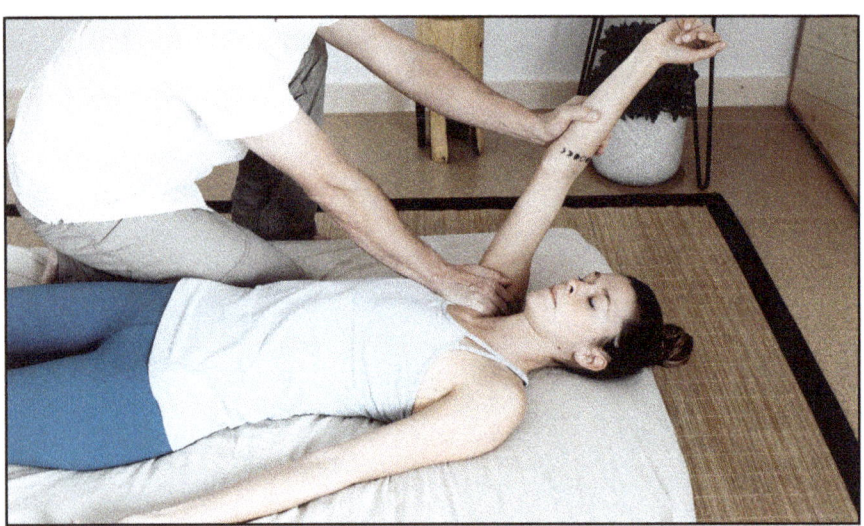

Thumb shiatsu hands: massage the hand on both sides, finishing by pulling the fingers softly to tips.

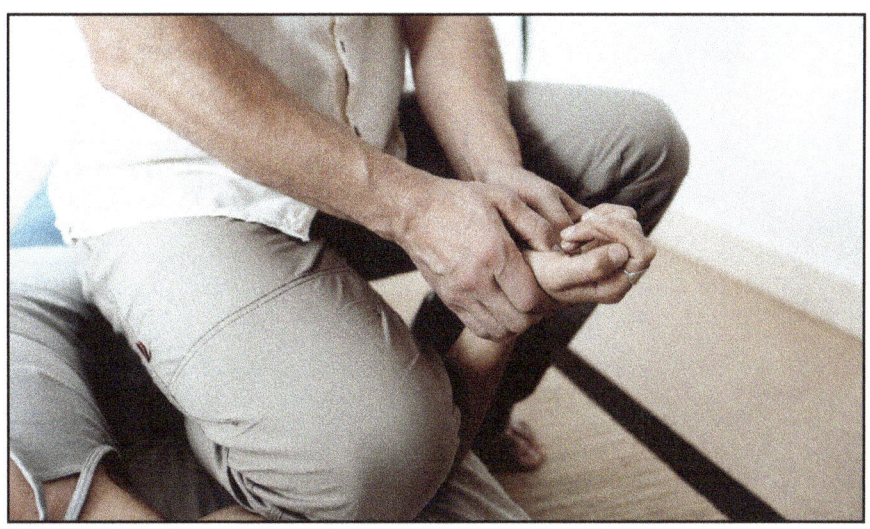

Stretching shiatsu arms: stretch the arm overhead from a standing position, pulling on each finger creating traction.

Repeat past sequence on both sides of the body.

Thumb shiatsu neck: hold the head in one hand and with the other hand massage the neck with the thumb from the spine to the side of the neck, from top to bottom, thumb shiatsu both sides of the neck.

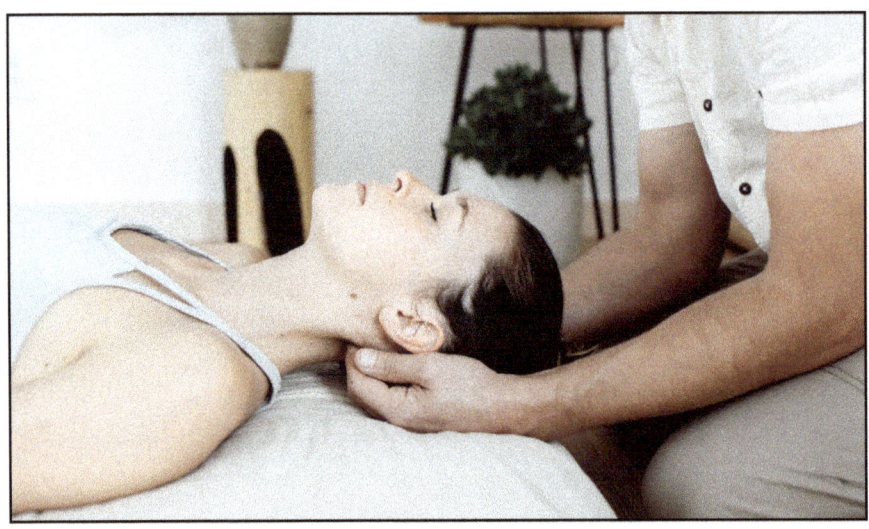

Thumb shiatsu forehead: hold point on forehead and sweep from center to temples over the forehead.

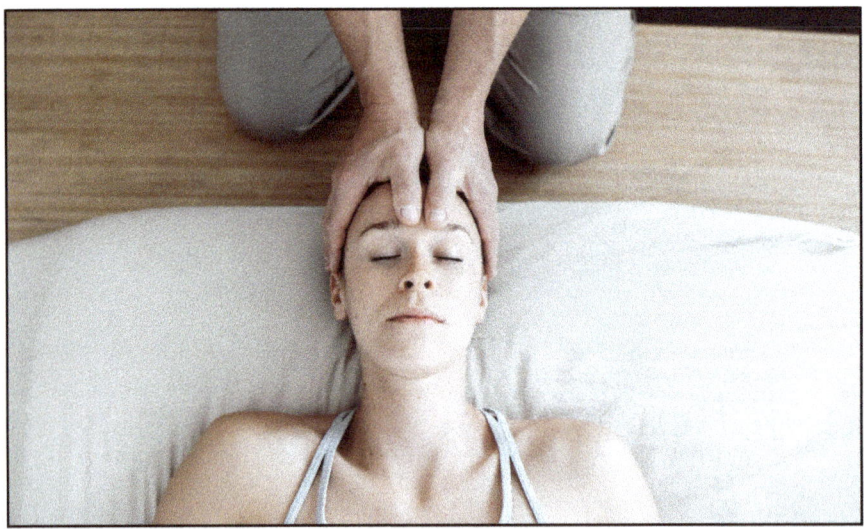

Thumb shiatsu head: press and hold top of head on pressure point for 1–2 seconds.

Palm shiatsu stomach: move back to abdomen, using palpations and slowly release your hand to end the session.

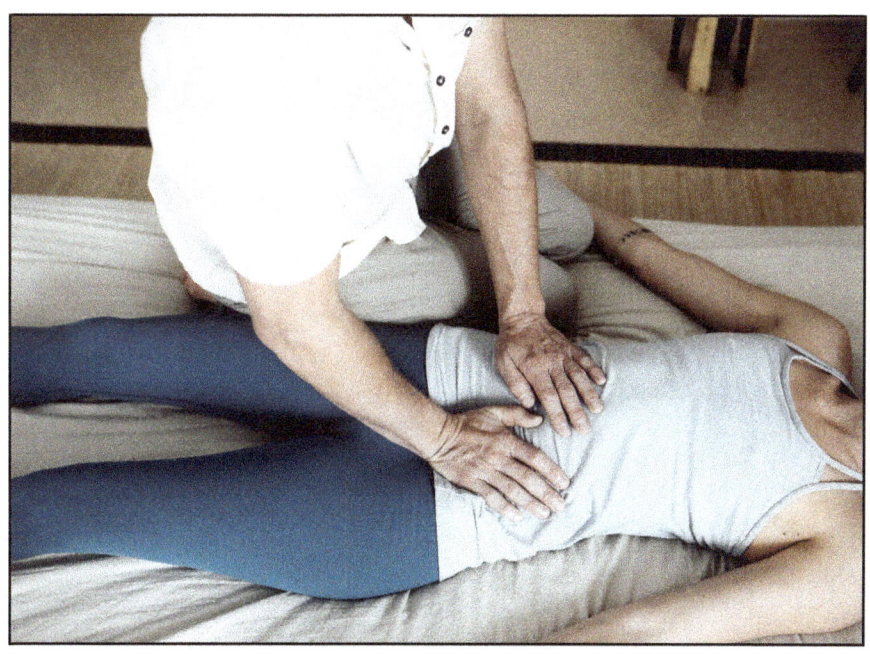

Appendix A — Channel Abbreviations

BL	Bladder
CV	Conception vessel
GB	Gallbladder
GV	Governing vessel
HE	Heart
KD	Kidney
LI	Large Intestine
LU	Lung
LV	Liver
P	Pericardium
SI	Small intestine
SP	Spleen
ST	Stomach
TW	Triple warmer

Appendix B — Meridian Chart

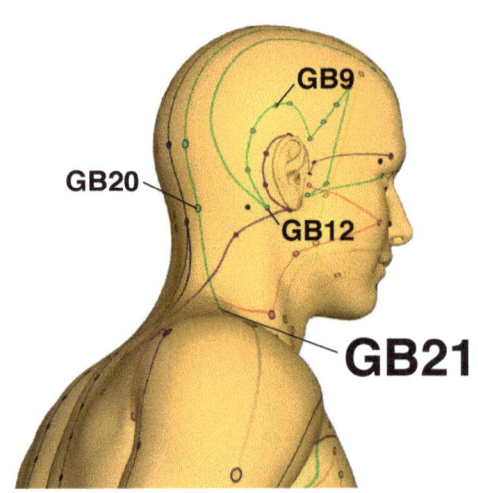

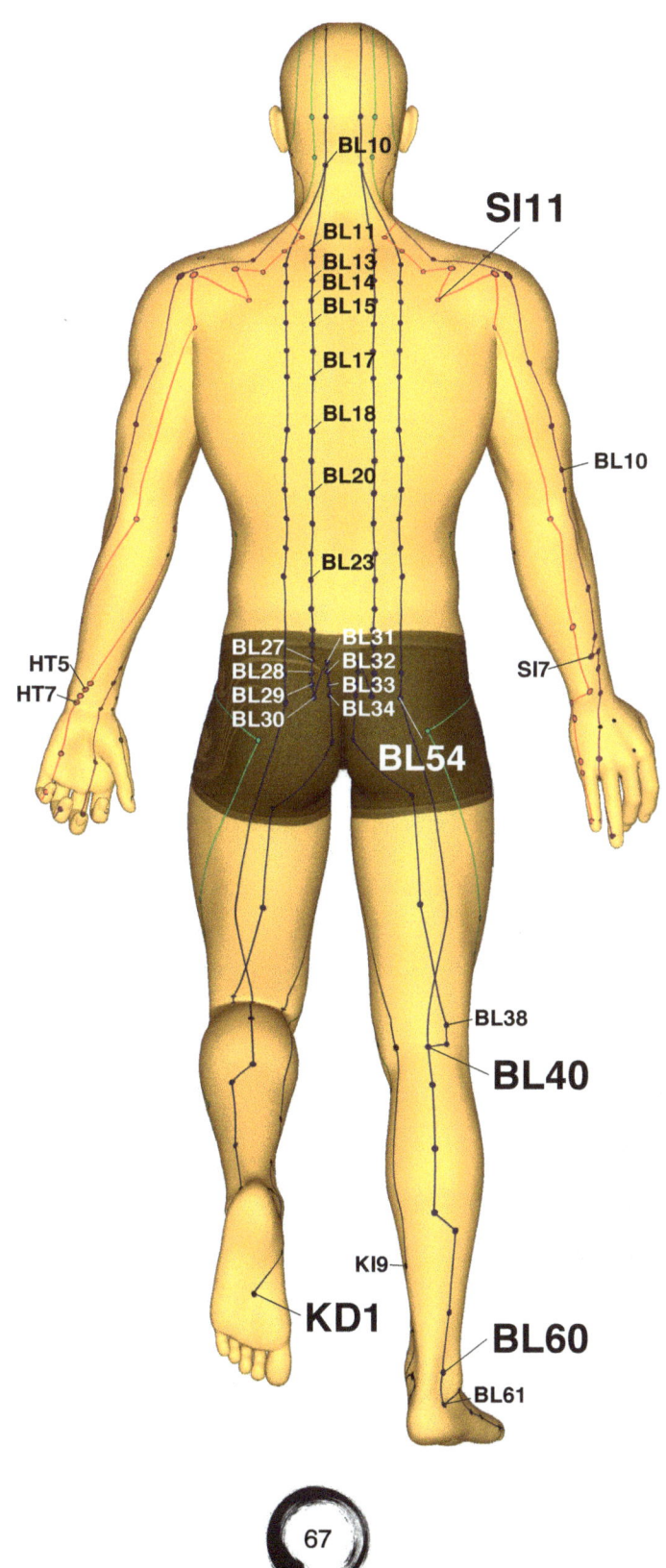

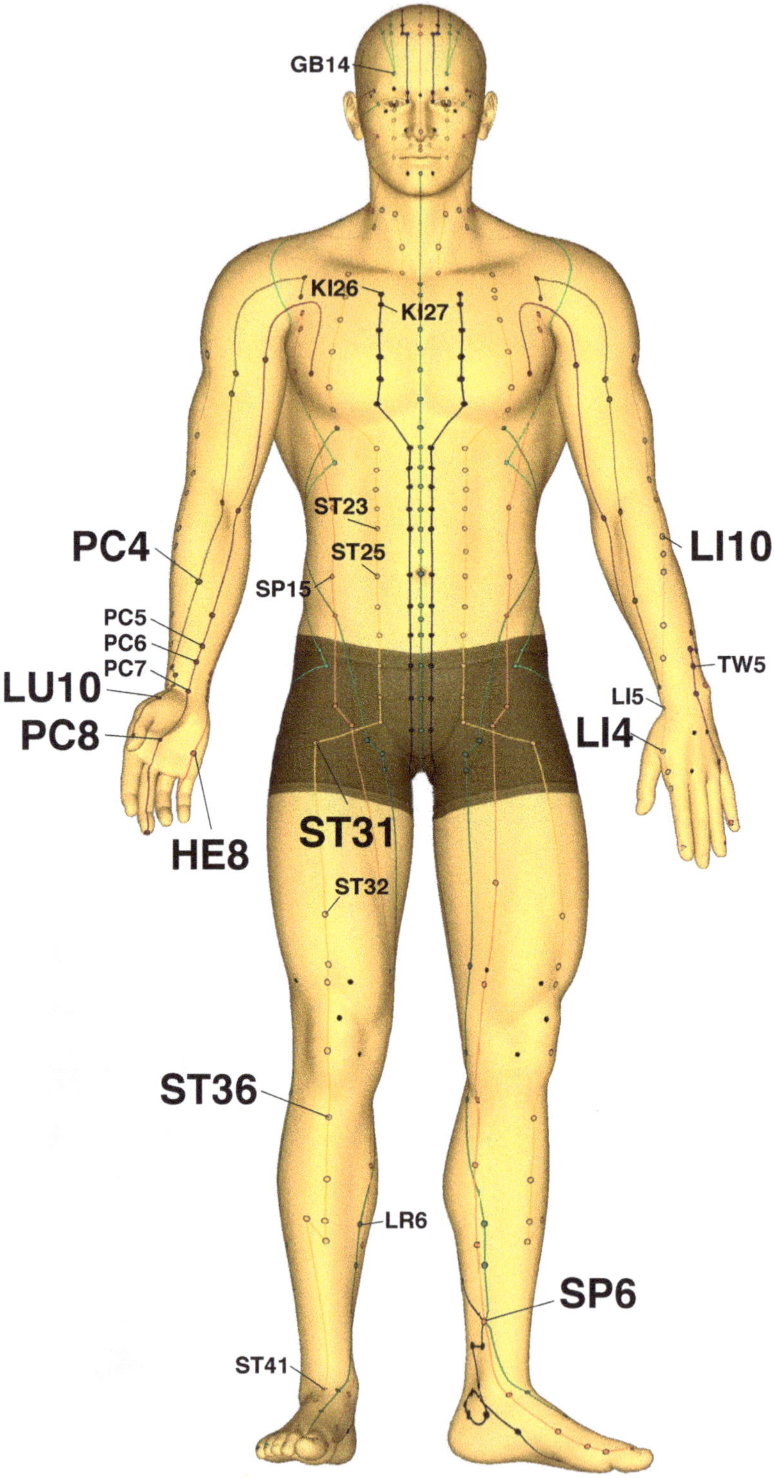
GB14
KI26
KI27
ST23
ST25
SP15
PC4
LI10
PC5
PC6
PC7
TW5
LU10
LI5
PC8
LI4
HE8
ST31
ST32
ST36
LR6
SP6
ST41

Glossary

Anma – Style of massage in Japan that evolved into Shiatsu, which originated from Tui Na, a massage technique that originated in China.

Benevolence – Kindness.

Brush off – Finishing a sequence by lightly brushing down the area that was just massaged. This technique is used when ending your work or when transitioning into the next sequence. Brushing off is used to remove old stirred up energy from your client's body and helps to relax and also to end the session.

Bubbling spring – Name of major body point Kidney 1. This is a point on the bottom of the foot used to both energize and calm a person at the same time.

Caring intention – Setting an intention before starting a massage is important in directing how you will perform the massage. To care about someone not only directs the physical but is also felt on the inside of one's body.

Cat shiatsu – While in a kneeling position, pushing down with both hands and alternating pressure from one hand to the other.

Channels – Channels are meridians in the body or energy pathways relating to the organ system.

Channel Abbreviations – the abbreviations are short hand for the organ system the channel is related to.

Conception Vessel – Major energy pathway running down the center of the front of the body's torso.

Elbow shiatsu – Using the point of the elbow for more direct pressure or the Ulna for more widespread pressure.

Fire Point – They are fire points on other meridians other than the fire meridian.

Forearm shiatsu – Similar to elbow shiatsu; using the blade of the forearm to press down, creating an even pressure.

Four finger shiatsu – Using the pads of all four fingers on one hand to directly diagnose the meridians and the Hara to treat imbalances.

Gluteus maximus – Muscle of the buttock.

Governing Vessel – Major energy pathway running down the center of the back of the body's torso.

Hara – The soft part of the belly or front of the torso where each organ system is represented. This is the area of the body where you can diagnosis and treat by palpating.

Heel or blade of hand shiatsu – Heel shiatsu uses the inside bottom of your hand just above your wrist to press on an area of the body. Blade shiatsu uses the outside edge of your hand to press on an area of the body.

Hold shiatsu – To stay on a pressure point for a set time allowing the nervous system to relax and release. The average time is two to four seconds (if needed, stay for up to 30 seconds). This method helps the client relax more deeply by switching them into their parasympathetic nervous system, also called the rest and digest mode.

Homeostasis – Point of balance in a system.

Illicit Crest – The curved border of the ilium bone.

ILLIUM bone – a bone located in the pelvis area just lateral of the sacrum.

IT Band – Iliotibial band which is a tendon that runs down the outer side of the legs.

Knee shiatsu – Using the area just below the knee with the support of your arms on the opposite side of the client's body to control the pressure.

The Lean – Off-setting your center of gravity allowing you to put more pressure on the client's body to work a pressure point or a hold.

Main gates – Major pressure points on the body.

Meridians – Energy pathways in the body, qi or life force, that flows through these body pathways to aid in assisting the performance of functions in the body.

Metacarpals – The five bones of the hand.

Mother hand – The process of holding one hand stationary on the client's body to help calm the nervous system and to sense the reaction of the body to the effects of pressure with your other hand.

Obscurity of the heat – when phlegm builds up in the heart organ it produces mental and physical pain also heat because of lack of flow of qi building up causing problems.

Old dog – Leaning on the client with a light supportive pressure with large parts of your body (such as the leg) to cover as much surface area as possible. The lean is equal pressure between you and your client, giving them the reassurance of trust. This method helps the acupressure point therapy penetrate deeper.

Open up the arms or legs – Refers to reestablishing natural qi flow in the channels or meridians.

Organs in Chinese medicine – Organs in Chinese medicine are functions in the body not the organs themselves.

Palm shiatsu – Using the whole palm of your hand to give soft, larger area pressure; this technique is used to understand what is going on in the body.

Palpation – Using your hand to examine or treat a specific area of the body.

Parasympathetic – Also known as the "Rest and Digest mode". This occurs when the nervous system is in relax mode letting the body heal, relax and digest.

Pent up – Extra amount of anything condensed in a specific area of the body.

Pericardium – Also known as the heart protector, it is a meridian reflecting the heart.

Phlegm – Extra water in the body that has condensed to a point, causing blockages and pain.

Post Traumatic Stress Disorder (PTSD) – Mental health state that occurs after someone goes through trauma, reliving the trauma through the onset of triggers in their present environment.

Pulse diagnosis – an in-depth look into someone's health through the pulse enabling one to feel any dysfunction in the body.

Qi - Bioelectrical energy in the body, pronounced "chee".

Rest and Digest mode – the parasympathetic nervous system of the body when it is at rest.

Rocking shiatsu – A method of applying pressure in the form of rocking back and forth continuously with your body. This method helps loosen the client's muscles as well as relax the nervous system.

Rotation shiatsu – Rotating the arm at the shoulder, elbow, and wrist; rotating the leg, knee, and foot.

Sacrum – The large triangle shaped bone at the bottom end of the spine, in the back of the hips.

Shiatsu – A type of massage originating in Japan using finger/elbow/knee pressure to aid the body in healing itself.

Shin – Front part of the leg just below the knee.

Squeeze and twist shiatsu – Alternate pressure between both hands with one hand staying in place while the other hand moves and presses down on a specific body part.

Supine – The body is facing up in a lying down posture on a surface.

Stretching shiatsu – Extending a body part to the full range of motion using slow and steady pressure.

Sympathetic – When the body is in "fight or flight mode", the nervous system shuts off non-critical functions in the body to direct the body to both a heightened awareness and more focused muscle functions.

Thumb and index muscle shiatsu – Your thumbs are supported against the bent index finger enabling you to press with both fingers at the same time.

Thumb shiatsu – Squeezing the thumb and hand together to provide pressure on a point in the body.

Trap – The trapezius muscle at the base of the neck used to tilt and turn the head, neck and arm.

Treatment hand – The hand that is moving and providing pressure to the area of the body being treated.

Triple Warmer – A meridian in charge of the three warmers of the body – upper, middle and lower body areas.

Tui Na – A style of massage that originated in China.

Ulna – A long bone found in the forearm area of the arm.

Warmers – Warmers is a term used to divide the body in three sections; upper, middle and lower. All three sections make up the body and help create life in the body as "warm equals life."

Wild energy – this is the qi energy that moves through the body which can cause illness and/or stress in the body.

Zen Shiatsu – A style of shiatsu that was invented by Master Shizuto Masunaga of Japan. Master Masunaga added more meridians in the body than standard shiatsu to expand the treatment and depth of healing.

Reference List

1. Atkinson,M., Touch for Children: Massage Reflexology, and Accupressure for Children from 4-12 years old. 2009, New York, NY: Octopus Books USA.

2. Benedek, D.M. and G.H. Wynn, Complementary and Alternative Medicine for PTSD. 2016, New York, NY: Oxford University Press.

3. Beresford-Cooke, C., Shiatsu Theory and Practice. Vol.3rd.2011, Edinburg: Churchill Livingstone.

4. Brady, L.H.,et al., The Effects of Shiatsu on Lower Back Pain. Journal of Holistic Nursing, 2001. 19(1): p.57-70.

5. Jarrett, Lonny S., The Clinical Practice of Chinese Medicine. Stockbridge, MA. Spirit Path Press, 2006.

6. Kessler, R.C., et al., Prevalence, Severity, and Comorbidity of a 12 month DSM-IV Disorders in the National Comorbidity Survey Replication. Archives of General Psychiatry, 2005. 62(6): p. 617-627.

7. Libby, D.J., C.E. Pilver, and R. Desai, Complementary and Alternative Medicine in VA Specialized PTSD Treatment Programs.
Psychiatric Services, 2012. 63(11): p. 1134-1136.

8. Long, A.F., The Effectiveness of Shiatsu: Findings from a Cross-European Prospective Observational Study. The Journal of Alternative and Complementary Medicine, 2008. 14(8): p. 921-930.

9. Robinson, N., A. Lorenc, and X. Liao, The Evidence for Shiatsu: A Systematic Review of Shiatsu and Acupressure. BMC Complementary and Alternative Medicine, 2011. 11(1): p.88

10. Schure, M.B., et al., Mindfulness-based Processes of Healing for Veterans With Post tramatic Stress Disorder. Journal of Complementary and Alternative Medicine,
In publication.

11. Tanielian, T. and L.H.E. Jaycox, Invisible Wounds of War. Summary and Recommendations for Addressing Psychological and Cognitive Injuries. 2008, RAND Corporation: Santa Monica, CA.

12. U.S. Census Bureau. American Community Survey, 2015 Data.

13. Weathers, F., et al., PCL_M for DSM-IV. 1991, National Center for PTSD – Behavioral Science Division.

Notes

Notes

Notes

Notes

Notes

Notes

Notes

www.ingramcontent.com/pod-product-compliance
Lightning Source LLC
Chambersburg PA
CBHW040518220526
45473CB00012B/2899